Structure Determination By Spectroscopic Methods

A Practical Approach

T0203950

Raul SanMartin

Department of Organic Chemistry II
Faculty of Science and Technology
University of the Basque Country (UPV/EHU)
Leioa (Bizkaia), Spain

María Teresa Herrero

Department of Organic Chemistry II
Faculty of Science and Technology
University of the Basque Country (UPV/EHU)
Leioa (Bizkaia), Spain

CRC Press
Taylor & Francis Group
Boca Raton   London   New York

CRC Press is an imprint of the
Taylor & Francis Group, an **informa** business

A SCIENCE PUBLISHERS BOOK

Cover credit: The authors have designed the cover themselves without using any copyrighted image.

CRC Press
Taylor & Francis Group
6000 Broken Sound Parkway NW, Suite 300
Boca Raton, FL 33487-2742

© 2021 by Taylor & Francis Group, LLC
CRC Press is an imprint of Taylor & Francis Group, an Informa business

No claim to original U.S. Government works

Version Date: 20200527

International Standard Book Number-13: 978-1-138-49158-8 (Hardback)

Library of Congress Cataloging-in-Publication Data

Names: SanMartin, Raul, 1968- author. | Herrero, María Teresa, 1974- author.
Title: Structure determination by spectroscopic methods : a practical approach / Raul SanMartin, María Teresa Herrero.
Description: Boca Raton : CRC Press, Taylor & Francis Group, [2020] | Includes bibliographical references and index.
Identifiers: LCCN 2020004497 | ISBN 9781138491588 (hardback)
Subjects: LCSH: Organic compounds--Spectra--Problems, exercises, etc. | Organic compounds--Structure--Problems, exercises, etc. | Spectrum analysis--Problems, exercises, etc.
Classification: LCC QC462.5 .S36 2020 | DDC 543/.5--dc23
LC record available at https://lccn.loc.gov/2020004497

Visit the Taylor & Francis Web site at
http://www.taylorandfrancis.com

and the CRC Press Web site at
http://www.routledge.com

Preface

The elucidation of the chemical structure of a pure compound is a problem solved by chemists on a daily basis. In this mission, spectrometric and spectroscopic techniques, such as infrared spectrophotometry, nuclear magnetic resonance spectroscopy, or mass spectrometry have become essential tools. However, although elucidation work has been greatly simplified by these methods, the participation of the researcher/student is still crucial, since only humans can rationalize the results by combining the information provided by the above spectroscopic procedures and propose a plausible structure. Students who want to familiarize themselves with structural determination must obviously have knowledge of organic structures as well as of the theory behind spectroscopic techniques. However, applying these theoretical concepts in real cases is the most effective training process to improve those necessary "human skills", and hence the core of this book, which is the step-by-step description of problem-solving strategies. The aim of the authors, who have taught courses in structural determination over years, is to guide the reader's learning. In this regard, the solutions to selected problems are extensively explained, including profuse explanations of how to analyze the spectroscopic data provided, the information that can be extracted from them, as well as the conclusions that can be reached. In addition to the initial problem spectra and spectral data, abundant figures and data tables support the text so that the reader can easily follow the explanations. Special attention has been given to nuclear magnetic resonance and mass spectrometry because nowadays these techniques alone can provide enough information for the determination of the structure of an unknown compound.

On the other hand, the student will find in this preface useful general information for structural determination, such as some bibliographic sources for spectroscopic methods and spectroscopic data, helpful databases and other on-line tools, general strategies and indispensable basic concepts such as the calculation of the molecular formula from elemental analysis or molecular mass and the degree of unsaturation, as well as a brief description of the provided spectra and summary tables of the most useful spectroscopic data.

Raul SanMartin
María Teresa Herrero

Acknowledgements

We would like to thank the individuals and institutions who have contributed to this book. Dr. Garazi Urgoitia helped us perform some of the experiments and provided useful advice on the degree of difficulty or suitability of some the problems proposed. Besides, several joint ventures with her were temporally postponed in order to finish the writing of most chapters. In this regard, we cannot but praise the patience of three of our PhD students, Aimar Garcia, Galder Llorente, and Maria Obieta, who probably felt that we were falling behind in some supervisory tasks. Of course, all the expertise in teaching structure determination would not be the same without the help of many students whose feedback improved the quality of the solutions.

All this work would have been impossible without the support of our families, who helped us cope with the extra workload associated with the preparation of this book. Sincerely, we owe you a lot, Aitor-Eki and Susana-June.

Finally, some of the experiments performed were supported by the Basque Government (IT-1405-19) and the Spanish Ministry of Economy and Competitiveness (CTQ2017-86630-P). Technical and human support provided by SGIker of UPV/EHU is gratefully acknowledged.

Contents

PROBLEMS

SOLUTIONS TO PROBLEMS

Bibliography

The user of this book should have fundamental knowledge of structure and bonding of organic compounds, as well as of spectroscopic and spectrometric techniques usually employed for structural determination. Nevertheless, the books and online resources listed below may be helpful when solving exercises or theoretical doubts. In this regard, the literature on structural determination is extensive, and the list below is just intended to recommend some selected sources.

Tables of Data and Databases

A knowledge of basic parameters, such as chemical shifts, coupling constants, stretching vibrations or fragmentation pathways is often instrumental for structure determination. The same applies to the comparison of experimental information with reference data. This is why the following information sources are so valuable:

- Pretsch, E., P. Bühlmann and M. Badertscher. 2010. Structure Determination of Organic Compounds: Tables of Spectral Data. Springer, Berlin.
- Gupta, R.R. and M.D. Lechner. 2006. Nuclear magnetic resonance data. Springer, Berlin.
- https://webbook.nist.gov/chemistry. Online data base of the National Institute of Standards and Technology of the U.S. Department of Commerce.
- https://bit.ly/2UdCvob. nmrshiftdb2 is a NMR database for organic structures. (Steinbeck et al. 2003).
- https://bit.ly/2CQy0Li. Spectral Database for Organic Compounds, SDBS, a free site organized by National Institute of Advanced Industrial Science and Technology (AIST) of Japan.
- http://www.nmrdb.org. This website created by The Institute of Chemical Sciences and Engineering (ISIC) of the Ecole Polytechnique Fédérale de Lausanne (EPFL) and the Universidad del Valle de Cali contains no NMR spectra database, but allows to predict ^{13}C and ^1H spectra thanks to the tool of the FCT-Universidade NOVA de Lisboa developed by Yuri Binev and Joao Aires-de-Sousa. (Binev et al. 2007, Banfi and Patiny 2008, Castillo et al. 2011, Aires-de-Sousa 2002).
- www.chemcalc.org. A useful online molecular formula calculator from accurate molecular mass. (Patiny and Borel 2013).

Structure Determination by Spectroscopic Methods

- Hesse, H., H. Meier and B. Zeels. 2007. Spectroscopic Methods in Organic Chemistry. Thieme Publishing Group. *Explanations of the spectroscopic methods usually employed in organic chemistry are given in this book. Besides, the reader will find many examples and helping tables.*
- Williams, D. and I. Fleming. 2008. Spectroscopic methods in organic chemistry. McGraw-Hill, London. *Guide to the interpretation of spectra for structure determination. The book discusses briefly the basics of the technics and it describes mainly how they work and how to read the spectra.*
- Pavia, D.L., G.M. Lampman, G.S. Kriz, J.R. Vyvyan. 2009. Introduction to Spectroscopy. Brooks/Cole, Cengage Learning, Belmont. *The book presents basic theoretical concepts of spectroscopy methods and several problems to introduce analysis of spectra.*
- Lambert, J.B., S. Gronert, H.F. Shurvell, A. Lightner. 2011. Organic Structural Spectroscopy. Peason Prentice Hall, New Jersey. *This book presents the fundamentals of the four main spectroscopic methods: nuclear magnetic resonance spectroscopy, mass spectrometry, infrared spectrophotometry, and ultraviolet-visible spectroscopy.*
- Silverstein, R.M., F.X. Webster and D.J. Kiemle. 2014. Spectrometric Identification of Organic Compounds. John Wiley & Sons. *The book is characterized by its problem-solving approach with reference charts and tables.*
- Sanders, J.K.M. and B.K. Hunter. 1994. Modern NMR Spectroscopy. A guide for Chemist. Oxford University Press, New York. *This book provides a non-mathematical, descriptive approach to modern NMR spectroscopy, taking examples from organic, inorganic, and biological chemistry as well as providing much practical advice about the acquisition and use of spectra. Spectra of readily available compounds illustrate each technique.*
- Jacobsen, N.E. 2007. NMR Spectroscopy Explained. Simplified Theory, Applications and Examples for Organic Chemist and Structural Biology. Jon Wiley & Sons, New Jersey. *This book attempts to go further in the understanding of the Nuclear Magnetic Resonance avoiding formal physics and quantum mechanics as much as possible.*
- Richards, S.A. and J.C. Hollerton. 2011. Essential practical NMR for organic chemistry. John Wiley & Sons. Singapore. *The book addresses some aspects of NMR theory but its focus is on data adquisition, problem solving and interpretation.*
- Gross, J.H. Mass Spectrometry. 2017. Springer, Switzerland. Third edition. *Advanced text that includes a complete chapter dedicated to the fragmentation of organic ions and interpretation of EI mass spectra.*
- Thompson, J.M. 2018. Mass Spectrometry. Pan Stanford Publishing Pte. Ltd., Singapore. *Introductory text with the basic theory and interpretative techniques of mass spectrometry. The book includes spectra of a wide range of organic compounds.*

Instrumentation

The spectra were recorded employing the following instruments:

- 500 MHz ^{1}H NMR and 125 MHz ^{13}C{^{1}H} NMR spectra and all 2D NMR experiments were obtained on a Bruker AV-500.
- 300 MHz ^{1}H NMR, 75.5 MHz ^{13}C{^{1}H} NMR and DEPT 135 spectra were obtained on a Bruker AC-300.
- Infrared spectra were recorded on a Jasco FT/IR–4100 spectrometer.
- Mass spectra were recorded employing a 70 eV electronic impact ionization source. GC-MS analyzes were carried out on a Hewlett Packard 5890II gas chromatograph with a 100% methyl polysiloxane capillary column (30 m × 0.25 mm × 0.25 μm) interfaced with an Hewlett Packard 5989B mass spectrometer. Helium was the carrier gas and an electron impact ionization voltage of 70 eV was used. HRMS spectra (EI or CI) were obtained from a Micromass® GCT Premier™ orthogonal acceleration time-of-flight (oa-TOF) GC mass spectrometer. ESI-MS spectra were recorded in positive ion mode using a Waters Acquity UPLC-MS QTOF system.

Spectral Data

Mass Spectra

In most cases, electronic impact mass spectra (EI-MS) or electrospray ionization-mass spectra (ESI-MS) are provided. Occasionally, the high resolution mass spectral molecular ion peak is given in order to calculate the molecular formula.

IR Data

The complete infrared spectrum (transmitance vs wavenumber) or, alternately, the main vibrational bands are given in most cases in order to help with the identification of functional groups.

1D-NMR Data

^{1}H NMR and ^{13}C NMR experiments are given in all the cases. Most compounds were obtained from commercials sources and were used without further purification. Expansion of complex multiplets is provided to aid in their analysis. Except for those problems in which the ^{1}H-^{13}C HSQC 2D-spectra are given, DEPT 135 is provided in order to determine the substitution of the carbon atoms.

2D-NMR Data

In the case of more challenging structures, 2D NMR experiments are provided. The ^{1}H-^{1}H COSY spectra correlation technique is used to identify spin-spin coupling correlations between protons. In this context, it is important to remember that although COSY experiments generally provide information regarding the coupling between protons through two and three bonds, the magnitude of the coupling constant is crucial when studying off-diagonal peaks. Bond-distances and bond-angles, substituents, and, mainly, dihedral angles between the coupled protons can considerably modify these constants. In addition, long-distance couplings are common in conjugated compounds and strained cyclic aliphatic systems.

On the other hand, edited HSQC spectra are used not only to identify proton-bearing carbons and to associate these carbons with their attached protons, but also to distinguish between CH_3, CH_2, and CH groups.

^{1}H-^{13}C HMBC experiments are provided to facilitate the connection of the different fragments of the molecule deduced by COSY experiments and the assignation of quaternary carbons. Finally, ^{1}H-^{1}H NOESY spectra are used in order to identify those nuclei that are close together in space.

Problem-Solving Strategies

There is no one valid way to carry out the spectroscopic analysis to determine the structure of an unknown organic compound. The key is to work methodically in order to obtain all the information present in the spectra, but optimizing the time required for analysis. Here are some tips for performing the task successfully. These tips have been implemented in the problem resolution.

1. Calculation of the molecular formula

Anyone familiar with chemistry knows that describing the structure of an organic molecule involves much more than knowing its molecular formula. The scientist must specify how the atoms are connected and how they are arranged in space. However, knowing the constituent elements of the compound and the number of atoms of each element present in the molecule in advance is a great advantage. Once the molecular formula is calculated, the rest of the data is employed to determine the atom connectivity and the spatial disposition.

Empirical formula can be obtained by performing an elemental analysis, also known as combustion analysis. Combustion analysis has been used since it was discovered by Lavoisier at the end of 18th century and provides quantitative information on the relative percentages of atom types. Being a combustion analysis, the percentage of oxygen is calculated by difference. These percentages are then divided by the atomic mass of the corresponding elements, and the corresponding results divided by the smallest one. If necessary, they are converted to whole numbers. The empirical formula is then written using the above numbers as subscripts, showing the smallest whole-number ratio of the different atoms in the compound. For example, if the percentages given by the elemental analysis are as follows: C: 15.54, H: 1.74, Br: 68.92, the calculation of the empirical formula of the compound $(C_3H_4Br_2O_2)_n$ should follow the steps below:

1. Percentages/atomic mass
   C: 15.54: 12.01 = 1.29
   H: 1.74: 1.008 = 1.73
   Br: 68.92: 79.90 = 0.86
   O: 13.80: 16.00 = 0.86
2. Results/smallest one
   C: 1.29: 0.86 = 1.5
   H: 1.73: 0.86 = 2.01

Br: 0.86: 0.86 = 1.0

O: 0.86: 0.86 = 1.0

3. Convert to whole numbers

C: $1.5 \times 2 = 3$

H: $2.01 \times 2 \approx 4$

Br: $1.0 \times 2 = 2$

O: $1.0 \times 2 = 2$

4. Write the empirical formula: $(C_3H_4Br_2O_2)_n$

In view of the empirical formula, the molecular formula of the compound is easily calculated if the molecular mass is known. If it were 231.87, the value of n would be 1 and the molecular formula $C_3H_4Br_2O_2$.

$$n = \frac{231.87}{(3 \times 12.01) + (4 \times 1.008) + (2 \times 79.90) + (2 \times 16.00)} = 1$$

Nowadays, high resolution mass spectrometry (HRMS) is commonly used in order to determine molecular formula. In fact, this is one of the most important applications of this technique. High-resolution instruments are able to determine the value of the m/z of an ion to the fourth or fifth decimal place. Such an accurate measurement of the m/z value for the molecular ion allows the determination of the molecular formula that fits the data. There are textbook tables and online formula calculators providing several options (Pavia et al. 2009, Silverstein et al. 2014, Patiny and Borel 2013).

If only a low-resolution mass spectrum is available (LRMS), the corresponding less accurate m/z value of the molecular ion can be used for an approximate calculation. The "rule of 13" comes in handy to identify the possible molecular formulas for a given mass (Bright and Chen 1983). This rule of 13 consists of the following steps: first, generation of a base formula C_nH_{n+r} of a hydrocarbon compound, where n and r are, respectively, the numerator and the remainder obtained by dividing the molecular mass by 13 (the mass of one carbon and one hydrogen). Second, in order to write a molecular formula including heteroatoms, we must subtract the mass of a combination of carbons and hydrogens that equals the masses of the elements to include (one carbon and three hydrogen in the case of the nitrogen, whose atomic mass is 15, one C and four H to insert an oxygen, whose atomic mass is 16, and so on).

As an example, for a compound with a molecular mass of 122 mass units, applying the rule of 13, the formula C_9H_{14} would be obtained, since $122 = (13 \times 9) + 5$. One carbon and four hydrogen atoms would then be subtracted in order to insert an oxygen and obtain the molecular formula of an organic compound fitting that molecular mass, so possible formulas would be $C_8H_{10}O$, $C_7H_6O_2$, $C_6H_4O_3$... Other formulas, such as

$C_8H_{12}N$ and C_7H_8NO, could also be considered. Sometimes, if there are not enough hydrogen atoms to subtract, we can subtract one carbon and add 12 hydrogen, or alternatively, we can add one carbon and subtract 12 hydrogen to obtain another formula. We should also keep in mind that if the molecular weight is even, the number of nitrogen atoms is 0 or an even number.

Anyway, it is clear that the number of possible formulas can be very high if only the molecular mass obtained by LRMS is available. Therefore, a more appropriate strategy is to take into account the data obtained by the spectroscopic techniques in order to limit the possibilities.

2. Calculation of the index of hydrogen deficiency, unsaturation index, or double-bond equivalents

The index of hydrogen deficiency is the number of π bonds and/or rings contained in a molecule and, therefore, it provides valuable information about the molecular structure. The unsaturation index can be calculated in different ways:

- The unsaturation index is the result of dividing the difference of hydrogen atoms between the experimental molecular formula and the formula for the corresponding hypothetical acyclic saturated compound (C_nH_{2n+2}) containing the same number of carbon atoms by two. For example, if the molecular formula of the compound were $C_3H_4Br_2O_2$, it would be C_3H_8 (C_nH_{2n+2}). Then, we must correct this formula for the heteroatoms present in the unknown. One hydrogen atom will be added for each Group V element present, and one hydrogen atom will be subtracted for each Group VII element. Accordingly, the formula would be C_3H_6. The difference in hydrogen atoms between C_3H_8 and C_3H_6 is 2, so the index of hydrogen deficiency is 1.
- Another more intuitive way is to consider the formula of a hydrocarbon equivalent to the unknown compound and compare it with the corresponding saturated hydrocarbon. To determine the formula for the hydrocarbon equivalent to the unknown compound, the heteroatoms must be replaced with hydrocarbon analogues. Hence, the formula is rewritten considering that a halogen atom (which forms one bond to fill its octet), an oxygen atom (which forms two bonds), and a nitrogen atom (which forms three bonds) are equivalent to hydrogen (–H), a methylene group (–CH_2–), and a methine group (=CH–), respectively. Silicon atoms are replaced with carbon atoms. In addition, if sulfur or phosphorus atoms are present in the compound, they will be considered as oxygen (S ~ O) and nitrogen (P ~ N) equivalents if their valence in the molecule is 2 in the case of sulfur, and 3 in the case of phosphorus. However, if

their valence is different, the calculation of the unsaturation index can be incorrect. Thus, sulfonate (R-SO$_3$H) and sulfate R-SO$_4$R') groups, with one and two π bonds respectively, do not contribute to the index of unsaturation because sulfur has 12 valence electrons in these cases.

Regarding the example we are considering, the hydrocarbon equivalent of C$_3$H$_4$Br$_2$O$_2$ would be C$_5$H$_{10}$, which has two hydrogen less than the corresponding saturated hydrocarbon (C$_5$H$_{12}$), so the unsaturation index is 1.

- There are also formulas available for calculating the double-bond equivalents. If the molecular formula was C$_c$H$_h$X$_x$N$_n$O$_o$ where X is a halogen atom, the index of hydrogen deficiency would be: UI = $c - (h/2) - (x/2) + (n/2) + 1$. So, for C$_3H_4Br_2O_2$, UI would be $3 - (4/2) - (2/2) + 1 = 1$. The formula only works with neutral molecules, and as mentioned before, it is not valid if the molecule contains other types of atoms, such as sulfur or phosphorus with higher oxidation states than oxygen or nitrogen.

3. *Check the presence of representative functional groups in the infrared spectrum*

Certain functional groups should be distinguished by a quick examination of the IR spectrum. It can be very useful to confirm the presence of hydroxyl or amine groups or carboxylic acids whose signals in ^{1}H NMR can be misleading. IR is also very helpful in the recognition of nitrile groups, which are not that recognizable by NMR.

4. *Examination of the NMR spectra*

The strategy may vary slightly depending on the number of experiments available to us. In general, two-dimensional nuclear magnetic resonance spectra will not be essential for the determination of structure of simple molecules, although they can be very valuable for the unambiguous assignment of the signals. By interpretation of NMR spectra, information is obtained at different levels, leading to a plausible structure. The main stages are described below:

- *Study of molecular symmetry*: In NMR, each nucleus has a resonance frequency dependent on the chemical environment. The comparison of the number of carbon signals appearing in ^{3}C{^{1}H} NMR with the number of carbon atoms from the molecular formula is essential to draw conclusions about the symmetry of the molecule. The more symmetric the molecule, the smaller the number of carbon signals, as there will be more carbon nuclei with an equivalent chemical environment that will resonate at the same frequency. In any case, we must not forget that the signals can overlap if the nuclei show similar chemical shifts.

- *Study of the chemical shifts. Determination of functional groups:* The information from the infrared spectrum can be complemented by studying the NMR chemical shifts. For example, the presence of nearby electronegative groups (separated by one to three bonds) causes a downfield shift of the signal due to the decrease of the electronic density around the nucleus. Another example, this time related to magnetic anisotropy, is the effect of multiple bonds or aromatic rings on the resonance frequency of nearby protons. A summary of the chemical shift ranges of the most common protons and carbons depending on the adjacent atoms and functional groups is shown in Figures 0.4 and Figure 0.6. More comprehensive series of empirical tables of chemical shifts of different nuclei as well as computer tools for their theoretical calculation are referenced in the bibliography section.
- *Determination of hydrocarbon structure fragments and spin systems:* The combination of ^{13}C NMR and DEPT experiments allow us to differentiate among primary, secondary, tertiary, and quaternary carbons. Heteronuclear correlation through one bond (HSQC, HMQC, HETCOR) provides C-H connectivity, which is key information in order to determine spin systems in the case of asymmetric molecules. Indeed, interpretation of ^{1}H NMR spectra of compounds containing stereogenic centers becomes more challenging due the signals of diastereotopic protons. Besides, the carbon dimension can be used to resolve the often overlapping proton dimensions.
  The splitting of the proton signals by the nearby active nuclei is a valuable source of information. When dealing with simple molecules, the analysis of multiplets, including measurement of the coupling constants, will be enough to deduce the spin systems. In more challenging problems, ^{1}H-^{1}H correlation data is essential. ^{1}H-^{1}H COSY experiments are a powerful tool that shows correlations due to J couplings so that conclusions on the spin systems can be easily drawn. As mentioned before, the examination of CH correlations prior to the analysis of the results from COSY is advisable if the signal overlapping is considerable.
- *Identification of the molecular skeleton:* Once the spin systems and the hydrocarbon fragments are mapped, it is time to put together all the pieces of the puzzle. Connecting the fragments can be obvious for small molecules, but it is challenging in the case of more complex molecules. In this stage, ^{13}C-^{1}H correlations through 2 or 3 bonds (or any other heteronuclear correlation) can be essential for finding the right connections between different spin systems or fragments that are not directly coupled (e.g., quaternary carbons or groups that show no coupling in ^{1}H NMR).

– Determination of the relative configuration or preferred conformation. Considering that ^{1}H-^{1}H coupling constants (J) are dependent to a large extent on the distance and the angle between coupled nuclei, they can provide information regarding relative configuration of stereogenic centers or favored conformations in rings or conformationally restricted molecules. Therefore, the analysis of well-resolved multiplets in ^{1}H NMR may provide clues about the relative disposition of the coupled nuclei.

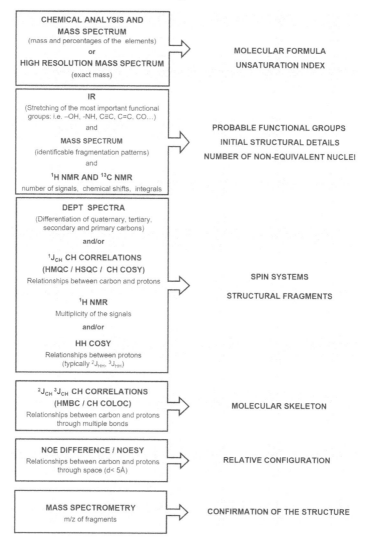

Figure 0.1   Suggested strategy to determine the structure of an unknown organic compound.

In this context, the experiments based on homonuclear NOE and rotating frame NOE effects (NOE difference, ^{1}H-^{1}H NOESY, ROESY) are very helpful, showing the coupling through space between nuclei that are close enough. This homonuclear effect is observed for protons separated by less than 5-6 Å.

5. Corroboration of the structure by mass spectrometry

Although it is true that finding out the structure of an unknown compound from only the mass spectrum is relatively rare, mass spectrometry is an essential tool for the identification of known structures due to the existence of large databases containing MS of several known compounds. In the context at hand, very helpful conclusions based on recognizable fragmentation patterns can be drawn from the analysis of EI-MS spectra. These conclusions can be used not only for a preliminary analysis of potential fragments or functional groups, but also for corroboration of the results.

The above-described strategy for the determination of the structure of an unknown organic compound is summarized in the previous figure (Fig. 0.1).

Summary Table of Some Useful Spectroscopic Data

Below are brief summaries of wavenumber values for vibration bands found in the infrared spectra of organic compounds, as well as chemical shift ranges and coupling constant values in proton and carbon-13 nuclear magnetic resonance (Prestch et al. 2010, Hesse et al. 2007, Pavia et al. 2009, Williams and Fleming 2008, Cookson et al. 1966, Cahill et al. 1969, Bothner-By 1965, Günther and Jikeli 1977, Günther 1992, Levitt 2001, Friebolin 2010, Levy et al. 1980, Marshall 1983).

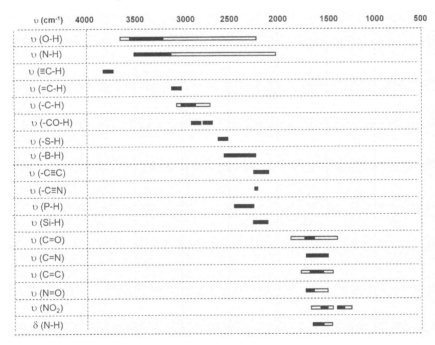

Figure 0.2 Summary table of the typical absorption bands in infrared spectroscopy.

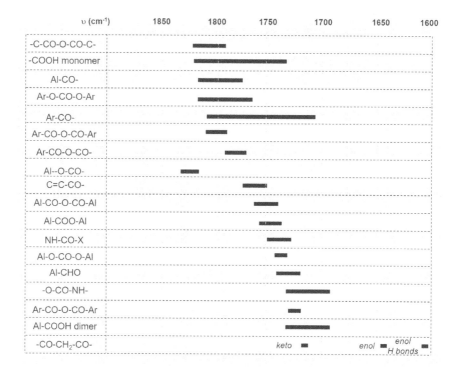

Figure 0.3 Summary table of carbonyl absorption bands in infrared spectroscopy.

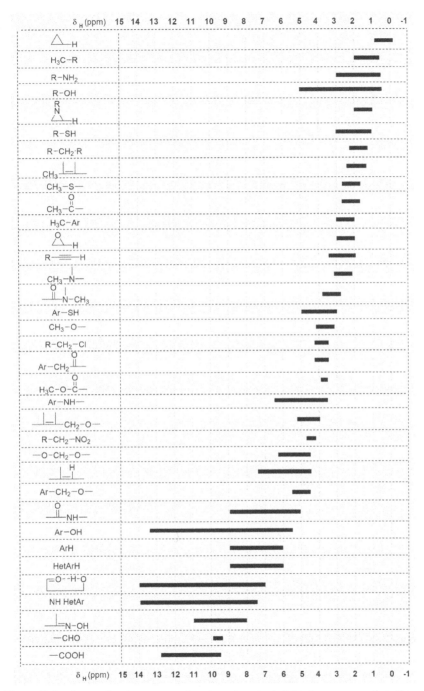

Figure 0.4   Approximate ranges of proton chemical shifts for common functional groups.

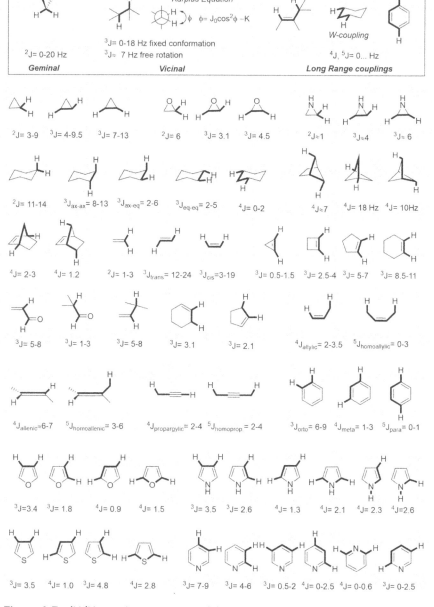

Figure 0.5 ¹H-¹H coupling constants (|J| in Hz) for a number of geminal, vicinal, and long-range couplings.

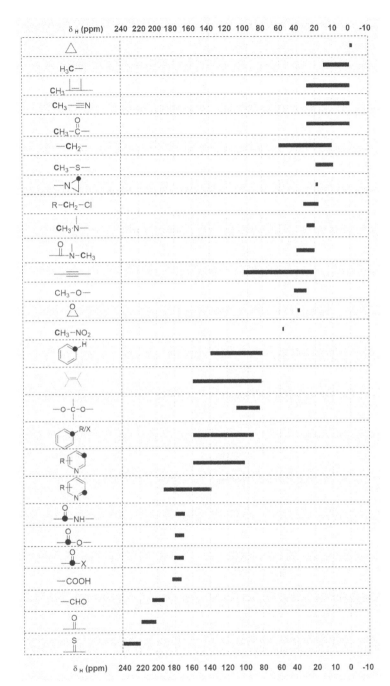

Figure 0.6 Approximate ranges of carbon chemical shifts for common functional groups.

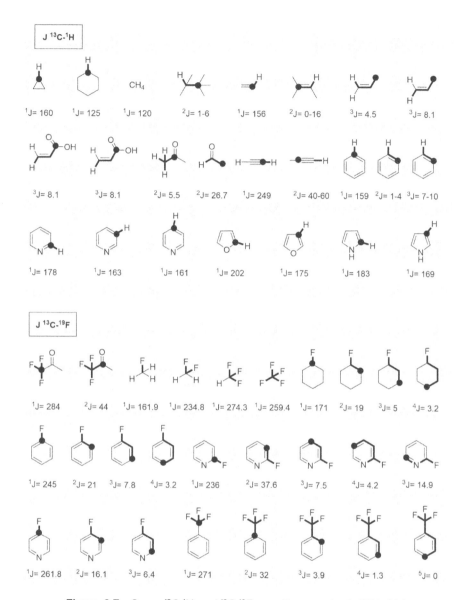

Figure 0.7   Some ^{13}C-^{1}H and ^{13}C-^{19}F coupling constants (|J| in Hz).

Abbreviations

Ar: Aromatic substituent
ax: Axial
bs: Broad singlet
$^{13}C\{^1H\}$ NMR: Broadband proton decoupled ^{13}C nuclear magnetic resonance
C_{arom}: Aromatic carbon
COSY: Correlation spectroscopy, two dimensional shift correlations
DEPT: Distortionless enhancement by polarization transfer
d: Doublet
dd: Doublet of doublets
ddd: Doublet of doublet of doublets
δ: Chemical shift (NMR) or bending vibration (IR)
EI: Electronic impact
ESI: Electrospray ionization
EDG: Electron donor group
endo: Endocyclic proton
eq: Equatorial
exo: Exocyclic proton
EWG: Electron withdrawing group
FT: Fourier transform
HMBC: Heteronuclear multiple bond correlation
HMQC: Heteronuclear multiple quantum coherence
HRMS: High resolution mass spectra
HSQC: Heteronuclear single quantum coherence
IR: Infrared spectra
J or ^1J, ^2J, ^3J, ^4J...: Spin-spin coupling constant through one, 2, 3, 4 bonds (one-bond coupling, germinal, vicinal, long-range coupling)
LRMS: Low resolution mass spectra
m: Multiplet or meta position on aromatic rings
M^+: Molecular ion
m/z: Mass-to-charge ratio
MS: Mass spectra (usually low resolution spectra)
NMR: Nuclear magnetic resonance
NOE: Nuclear Overhauser effect
NOESY: Nuclear Overhauser effect spectroscopy
pg: Page

p: Para position on aromatic rings
o: Orto position on an aromatic ring
oct: Octect
ppm: Parts per million
qC: Fully substituted carbon
s: Singlet
sept: Septet
sxt: Sextet
tC: Tertiary carbon
t: Triplet
υ: Stretching vibration
q: Quartet
qui: Quintet
ψquint: Apparent quintet

PROBLEMS

Problem 1

A colorless liquid was isolated as the main product from the halogenation of a mixture of alkanes. Determine its structure from the following spectra and spectral data.

EI-MS (70 eV)

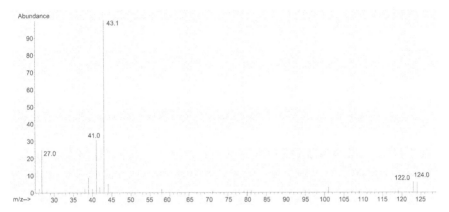

Figure 1.1 EI-MS unknown compound **1**.

IR SPECTRUM (Thin film)

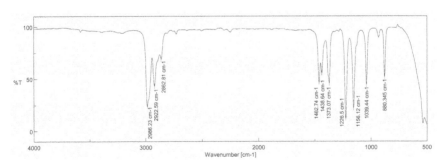

Figure 1.2 IR spectrum of unknown compound **1**.

¹H NMR SPECTRUM (300 MHz, CDCl₃)

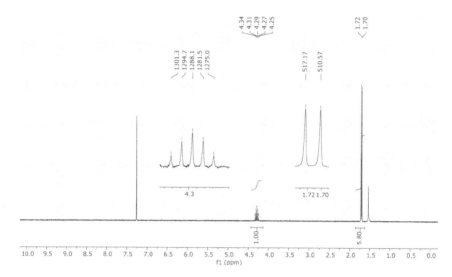

Figure 1.3   ¹H NMR spectrum of unknown compound **1**.

¹³C{¹H} NMR AND DEPT-135 SPECTRA (75.5 MHz, CDCl₃)

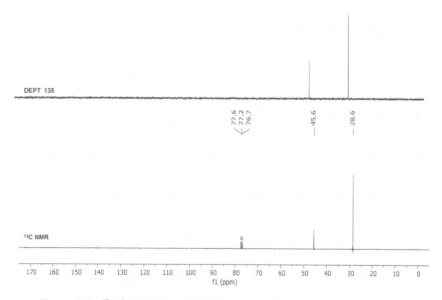

Figure 1.4   ¹³C{¹H} NMR and DEPT spectra of unknown compound **1**.

Problem 2

Catalytic hydrogenation of a sample of coal tar provided, after chromatographic separation, a yellowish white powder. Draw a structure for this compound that is consistent with all the spectra and spectral data presented below.

HRMS: (M⁺) found 180.0939

EI-MS (70 eV)

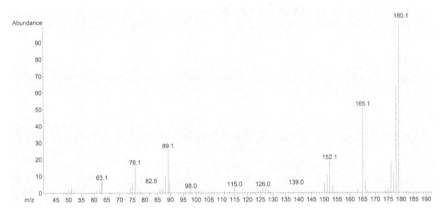

Figure 2.1 EI-MS of unknown compound **2**.

TABLE 2.1 Main peaks found in the EI-MS spectrum of unknown compound **2**.

m/z	Abundance (%)	m/z	Abundance (%)	m/z	Abundance (%)
75.10	6.0	139.10	6.0	176.10	29.0
76.10	16.0	150.10	9.0	177.20	18.0
82.60	5.0	151.10	17.0	178.20	18.0
88.10	10.0	152.10	29.0	179.20	96.0
89.10	29.0	153.10	6.0	180.20	100.0
89.90	8.0	165.10	81.0	181.20	35.0
115.10	5.0	166.10	12.0	182.10	3.0

IR SPECTRUM (Thin film)

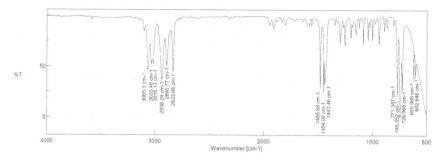

Figure 2.2   IR spectrum of unknown compound **2**.

¹H NMR SPECTRUM (300 MHz, CDCl₃)

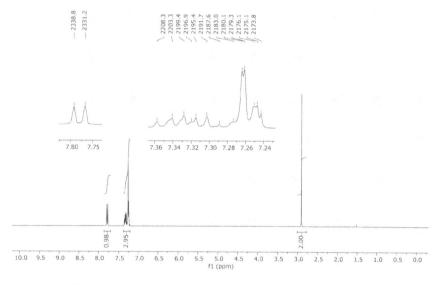

Figure 2.3   ¹H NMR spectrum of unknown compound **2**.

TABLE 2.2   ¹H NMR peaks of unknown compound **2**.

δ (ppm)	υ (Hz)	δ (ppm)	υ (Hz)	δ (ppm)	υ (Hz)	δ (ppm)	υ (Hz)
7.79	2338.8	7.33	2199.4	7.29	2187.6	7.25	2176.1
7.77	2331.2	7.32	2196.9	7.27	2183.0	7.25	2175.1
7.36	2208.3	7.31	2195.4	7.26	2180.1	7.24	2173.8
7.34	2203.3	7.30	2191.7	7.26	2179.3	2.90	871.3

¹³C{¹H} NMR AND DEPT-135 SPECTRA (75.5 MHz, CDCl₃)

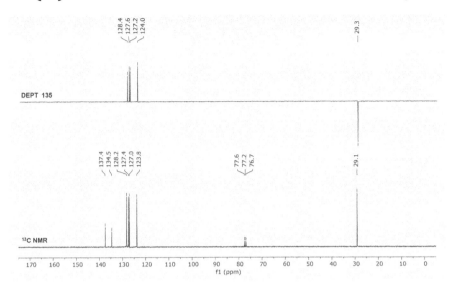

Figure 2.4 ¹³C{¹H} NMR and DEPT spectra of unknown compound **2**.

Problem 3

A pale yellow liquid with a fishy odor was isolated by acid-base extraction of a commercially available shampoo. Deduce its identity from the following spectra.

ESI-EM

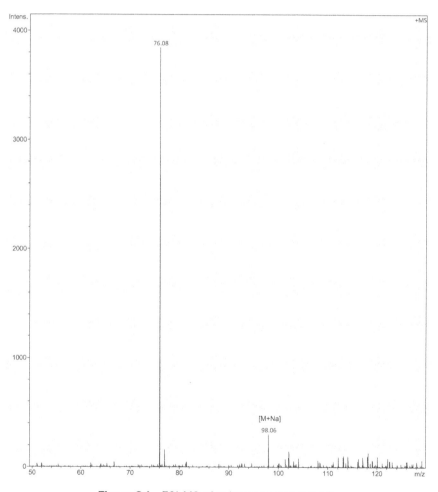

Figure 3.1 ESI-MS of unknown compound **3**.

EI-MS (70 eV): m/z (%): 18.0 (20), 28 (11.2), 29.0 (5.1), 30 (100.0), 31.0 (7.3), 44.0 (7.8), 56 (4.2), 57.0 (2.5)

IR SPECTRUM (Thin film)

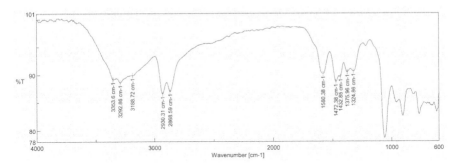

Figure 3.2 IR spectrum of unknown compound **3**.

¹H NMR SPECTRUM (300 MHz, CDCl₃)

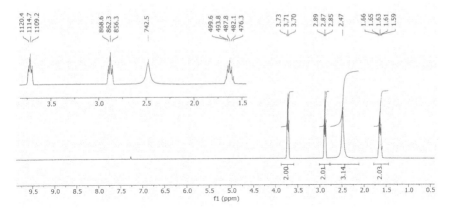

Figure 3.3 ¹H NMR spectrum of unknown compound **3**.

¹³C{¹H} NMR AND DEPT SPECTRA (75.5 MHz, CDCl₃)

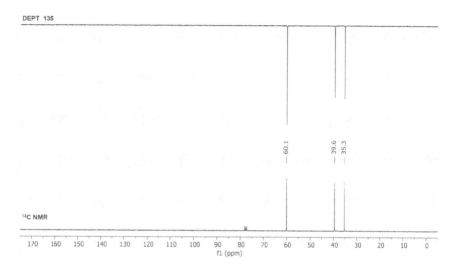

Figure 3.4   ¹³C{¹H} NMR and DEPT spectra of unknown compound **3**.

Problem 4

A chromatographically pure colorless oil was analyzed by several spectroscopic techniques. Construct a plausible structure for this unknown compound.

EI-MS (70 eV)

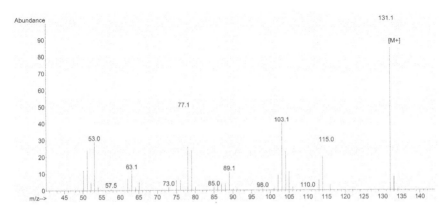

Figure 4.1 EI-MS of unknown compound **4**.

TABLE 4.1 Main peaks found in EI-MS of unknown compound **4**.

m/z	Abundance (%)	m/z	Abundance (%)	m/z	Abundance (%)
50.10	12.0	79.10	24.0	115.00	28.0
51.10	24.0	89.10	11.0	131.10	100.0
53.00	29.0	103.10	40.0	132.00	85.0
63.10	12.0	104.10	23.0	133.90	8.0
77.10	49.0	105.00	11.0		
78.10	26.0	114.00	20.0		

IR SPECTRUM (Thin film): υ (cm^{-1}) 3414.3 (broad), 3290.9, 3064.3, 2985.3, 2905.2, 2118.6, 1956.2, 1701.1, 1669.0, 1651.2, 1496.3, 1451.2, 948.4, 739.8 and 699.5

¹H NMR SPECTRUM (300 MHz, CDCl₃)

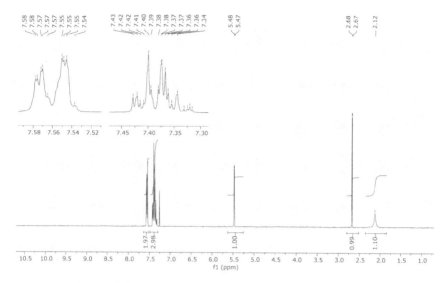

Figure 4.2   ¹H NMR spectrum of unknown compound **4**.

TABLE 4.2   ¹H NMR peaks of unknown compound **4**.

δ (ppm)	υ (Hz)	δ (ppm)	υ (Hz)	δ (ppm)	υ (Hz)	δ (ppm)	υ (Hz)
7.58	2274.3	7.54	2262.2	7.38	2213.6	7.32	2197.2
7.58	2273.8	7.43	2229.7	7.37	2213.1	7.32	2195.8
7.57	2272.7	7.42	2227.4	7.37	2211.2	7.26	2178.8
7.57	2272.1	7.42	2225.6	7.36	2209.6	5.48	1643.6
7.57	2270.5	7.41	2223.5	7.36	2207.5	5.47	1641.4
7.55	2266.1	7.40	2220.9	7.34	2204.2	2.68	803.9
7.55	2265.5	7.39	2219.4	7.33	2200.4	2.67	801.6
7.55	2264.7	7.38	2215.2	7.33	2198.5	2.12	637.1

¹³C{¹H} NMR AND DEPT-135 SPECTRA (75.5 MHz, CDCl₃)

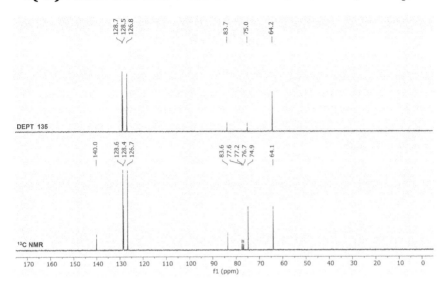

Figure 4.3 ¹³C{¹H} NMR and DEPT spectra of unknown compound **4**.

Problem 5

Propose a structure from the following spectra and spectral data. The compound is found as light yellow prisms fairly soluble in methanol.

EI-MS (70 eV)

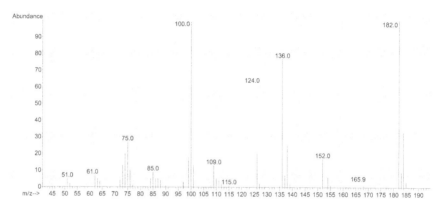

Figure 5.1   EI-MS of unknown compound **5**.

TABLE 5.1   Main peaks in EI-MS of unknown compound **5**.

m/z	Abundance (%)	m/z	Abundance (%)	m/z	Abundance (%)
73.00	13.0	101.00	13.0	152.00	17.0
74.00	20.0	109.00	13.0	182.00	100.0
75.00	27.0	124.00	62.0	183.00	9.0
76.00	10.0	126.00	20.0	183.90	33.0
99.00	18.0	136.00	77.0	184.90	3.0
100.00	99.0	138.00	25.0		

IR SPECTRUM (Thin film)

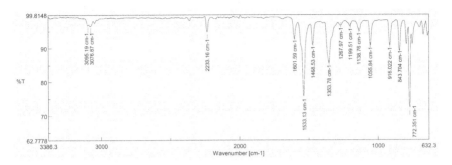

Figure 5.2 IR spectrum of unknown compound **5**.

¹H NMR SPECTRUM (300 MHz, CDCl₃)

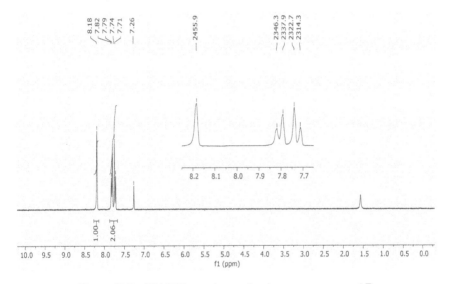

Figure 5.3 ¹H NMR spectrum of unknown compound **5**.

¹³C{¹H} NMR AND DEPT-135 SPECTRA (75.5 MHz, CDCl₃)

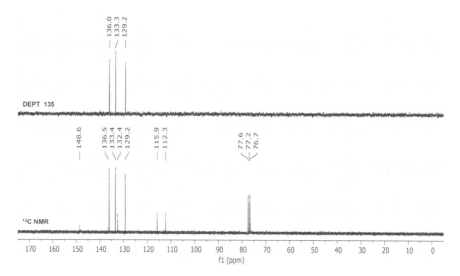

Figure 5.4   ¹³C{¹H} NMR and DEPT spectra of unknown compound **5**.

Problem 6

Construct a plausible structure for the unknown degradation product from lignin given its spectra and spectral data below.

EI-MS (70 eV)

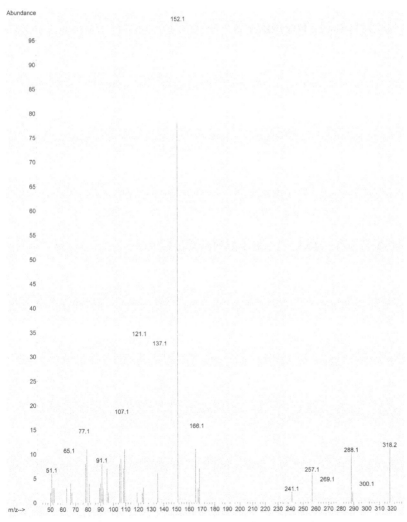

Figure 6.1 EI-MS of unknown compound **6**.

TABLE 6.1   Main peaks in EI-MS of unknown compound **6**.

m/z	Abundance (%)	m/z	Abundance (%)	m/z	Abundance (%)	m/z	Abundance (%)
51.10	6.0	95.10	7.0	122.10	6.0	165.10	11.0
65.10	10.0	105.00	8.0	135.10	6.0	166.10	15.0
77.10	14.0	106.10	9.0	137.10	32.0	257.10	6.0
78.10	8.0	107.10	18.0	139.10	8.0	288.10	10.0
79.10	11.0	108.10	7.0	151.10	78.0	318.10	11.0
91.10	8.0	109.10	11.0	152.10	100.0	319.20	2.0
94.00	5.0	121.10	34.0	153.10	11.0		

IR SPECTRUM (Thin film)

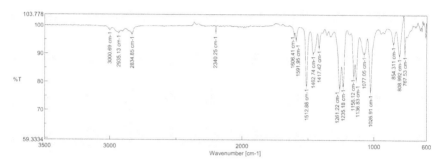

Figure 6.2   IR spectrum of unknown compound **6**.

¹H NMR SPECTRUM (300 MHz, CDCl₃)

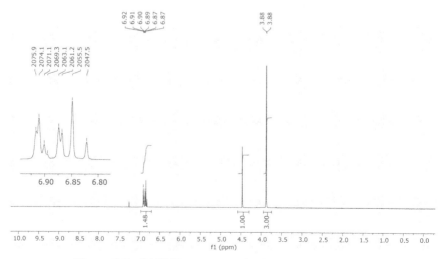

Figure 6.3   ¹H NMR spectrum of unknown compound **6**.

¹³C{¹H} NMR AND DEPT SPECTRA (75.5 MHz, CDCl₃)

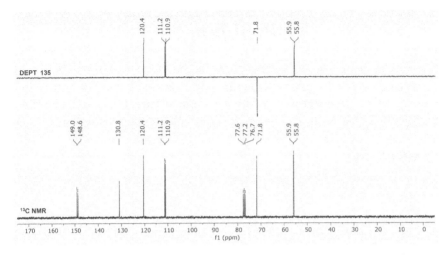

Figure 6.4 ¹³C{¹H} NMR and DEPT spectra of unknown compound **6**.

Problem 7

Some pale yellow crystals with a musty odor, isolated from tea and coffee, were analyzed in order to determine their structure. Deduce the identity of this food additive from the following spectra and spectral data.

EI-MS (70 eV)

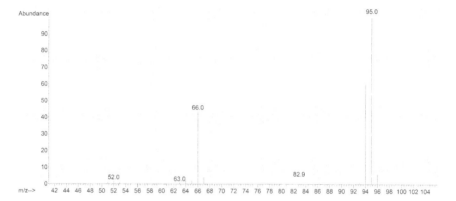

Figure 7.1 EI-MS of unknown compound **7**.

TABLE 7.1 Main peaks in EI-MS of unknown compound **7**.

m/z	Abundance (%)	m/z	Abundance (%)
51.00	1.0	67.00	4.0
52.00	2.0	82.90	4.0
53.00	1.0	84.90	3.0
63.00	1.0	94.00	59.0
64.00	2.0	95.00	100.0
65.00	2.0	96.00	6.0
66.00	42.0		

Problems

IR SPECTRUM (Thin film)

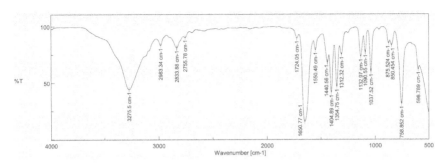

Figure 7.2 IR spectrum of unknown compound **7**.

¹H NMR SPECTRUM (300 MHz, CDCl₃)

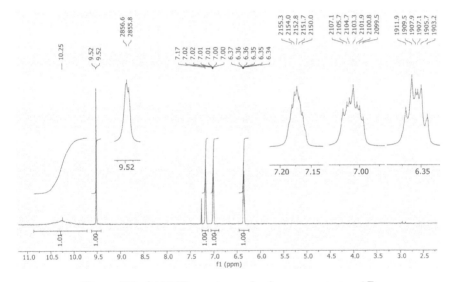

Figure 7.3 ¹H NMR spectrum of unknown compound **7**.

¹³C{¹H} NMR AND DEPT SPECTRA (75.5 MHz, CDCl₃)

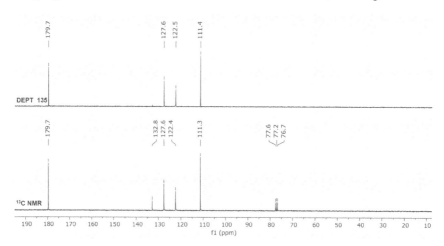

Figure 7.4   ¹³C{¹H} NMR and DEPT spectra of unknown compound **7**.

Problem 8

Determine the structure of the cross-coupling reagent whose spectra are given below.

EI-MS (70 eV)

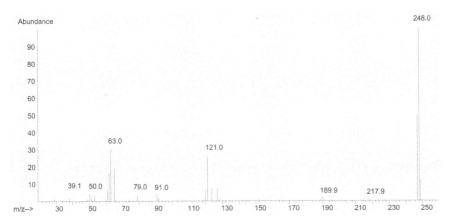

Figure 8.1   EI-MS of unknown compound **8**.

IR SPECTRUM (Thin film)

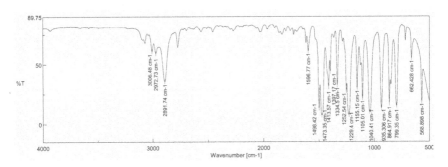

Figure 8.2   IR spectrum of unknown compound **8**.

¹H NMR SPECTRUM (300 MHz, CDCl₃)

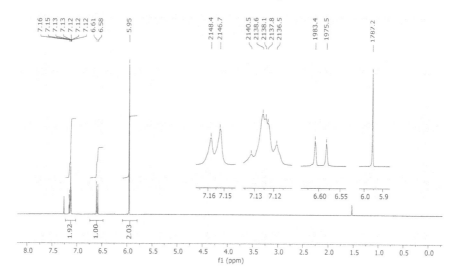

Figure 8.3   ¹H NMR spectrum of unknown compound **8**.

¹³C{¹H} NMR AND DEPT SPECTRA (75.5 MHz, CDCl₃)

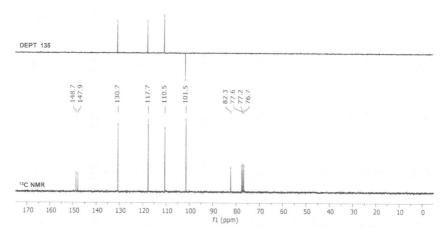

Figure 8.4   ¹³C{¹H} NMR and DEPT spectra of unknown compound **8**.

Problem 9

Construct a plausible structure for the unknown compound whose spectra and spectral data are presented below. It is an important raw material for copolymerization reactions.

EI-MS (70 eV)

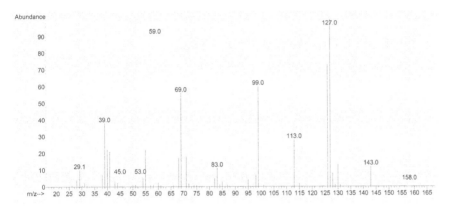

Figure 9.1 EI-MS of unknown compound **9**.

TABLE 9.1 Main peaks in EI-MS of unknown compound **9**.

m/z	Abundance (%)	m/z	Abundance (%)	m/z	Abundance (%)
29.10	10.0	67.00	12.0	113.00	28.0
39.00	38.0	68.00	17.0	126.00	73.0
40.00	22.0	69.00	56.0	127.00	100.0
41.00	21.0	71.00	18.0	130.00	13.0
55.00	22.0	83.00	11.0	143.00	12.0
59.00	91.0	99.00	60.0	158.00	3.0

IR SPECTRUM (Thin film)

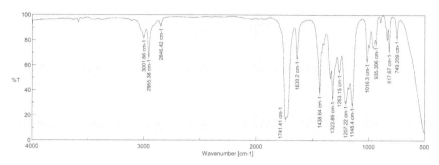

Figure 9.2   IR spectrum of unknown compound **9**.

¹H NMR SPECTRUM (300 MHz, CDCl₃)

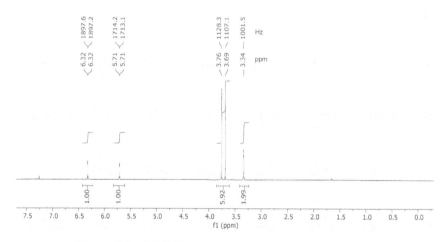

Figure 9.3   ¹H NMR spectrum of unknown compound **9**.

¹³C{¹H} NMR AND DEPT SPECTRA (75.5 MHz, CDCl₃)

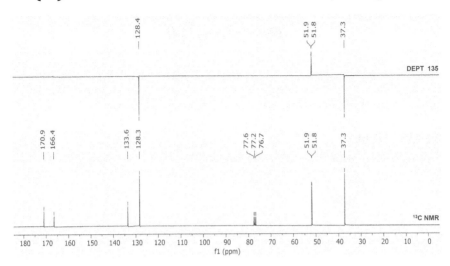

Figure 9.4 ¹³C{¹H} NMR and DEPT spectra of unknown compound **9**.

Problem 10

Deduce the identity of a light yellow liquid, commonly used as a flavoring agent, from the spectra and spectral data provided below.

EI-MS (70 eV)

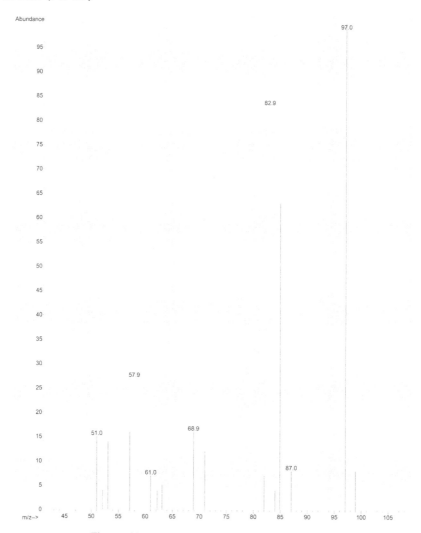

Figure 10.1 EI-MS of unknown compound **10**.

IR SPECTRUM (Thin film)

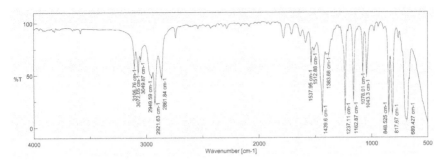

Figure 10.2 IR spectrum of unknown compound **10**.

¹H NMR SPECTRUM (300 MHz, CDCl₃)

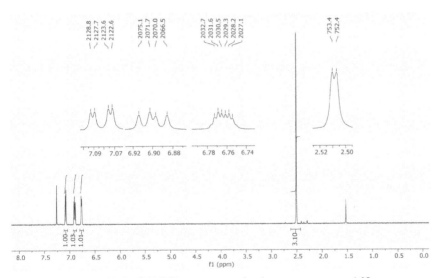

Figure 10.3 ¹H NMR spectrum of unknown compound **10**.

TABLE 10.1 ¹H NMR peaks of unknown compound **10**.

δ (ppm)	υ (Hz)	δ (ppm)	υ (Hz)	δ (ppm)	υ (Hz)
7.09	2128.8	6.90	2070.0	6.76	2028.2
7.09	2127.7	6.89	2066.5	6.75	2027.1
7.08	2123.6	6.77	2032.7	2.51	753.4
7.07	2122.6	6.77	2031.6	2.51	752.4
6.91	2075.1	6.77	2030.5		
6.90	2071.7	6.76	2029.3		

¹³C{¹H} NMR AND DEPT SPECTRA (75.5 MHz, CDCl₃)

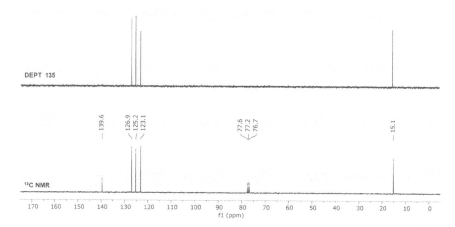

Figure 10.4   ¹³C{¹H} NMR and DEPT spectra of unknown compound **10**.

Problem 11

Determine the structure of the food flavoring whose spectra and spectral data are shown below.

IR SPECTRUM (Thin film)

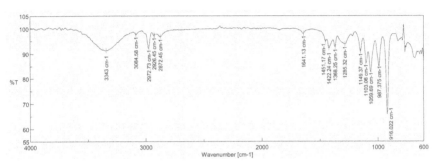

Figure 11.1 IR spectrum of unknown compound **11**.

¹H NMR SPECTRUM (300 MHz, CDCl₃)

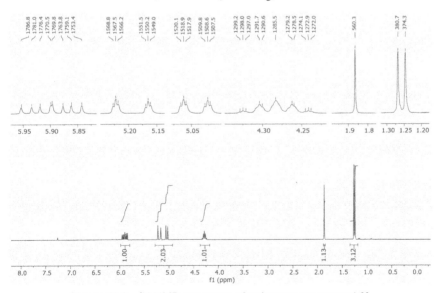

Figure 11.2 ¹H NMR spectrum of unknown compound **11**.

TABLE 11.1   ¹H NMR peaks of unknown compound **11**.

δ (ppm)	υ (Hz)	δ (ppm)	υ (Hz)	δ (ppm)	υ (Hz)
5.95	1786.8	5.17	1551.5	4.32	1298.0
5.93	1781.0	5.17	1550.2	4.32	1297.0
5.92	1776.4	5.16	1549.0	4.30	1291.7
5.90	1770.5	5.06	1520.1	4.30	1290.6
5.90	1769.8	5.06	1518.9	4.28	1285.5
5.88	1763.8	5.06	1517.9	4.26	1279.2
5.86	1759.1	5.03	1509.8	4.26	1278.5
5.84	1753.4	5.03	1508.6	4.25	1274.1
5.23	1568.8	5.02	1507.5	4.24	1272.9
5.22	1567.5	4.33	1299.2	4.24	1272.0
5.22	1566.2				

¹³C{¹H} NMR AND DEPT SPECTRA (75.5 MHz, CDCl₃)

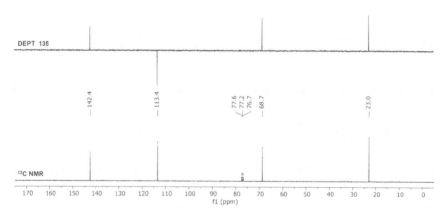

Figure 11.3   ¹³C{¹H} NMR and DEPT spectra of unknown compound **11**.

HRMS: m/z = 72.0571

EI-MS (75 eV): (m/z, %): 72.0 (2.6), 71.0 (13.8), 57.0 (100.0), 55.0 (10.3), 45.0 (30.6), 43 (73.1), 39 (13.7), 31.0 (10.4), 29.0 (24.4), 27.0 (29.9)

Problem 12

What is the structure of the oily substance with a sweet nutty odor, given its spectra?

EI-MS (70 eV)

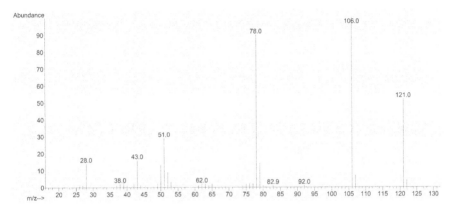

Figure 12.1   EI-MS of unknown compound **12**.

IR SPECTRUM (Thin film)

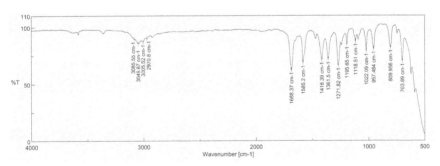

Figure 12.2   IR spectrum of unknown compound **12**.

¹H NMR SPECTRUM (300 MHz, CDCl₃)

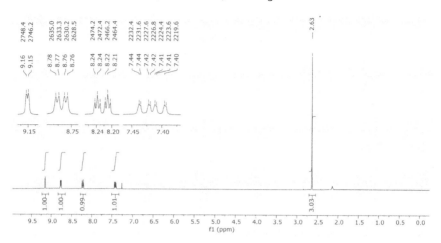

Figure 12.3   ¹H NMR spectrum of unknown compound **12**.

¹³C{¹H} NMR AND DEPT SPECTRA (75.5 MHz, CDCl₃)

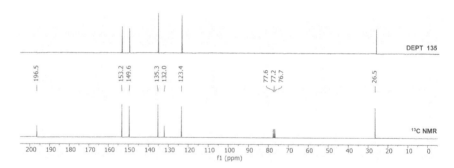

Figure 12.4   ¹³C{¹H} NMR and DEPT spectra of unknown compound **12**.

Problem 13

A commonly used alkylating agent was analyzed by spectroscopic methods in order to deduce its identity. Construct a plausible structure and assign as many signals as possible from the following spectra.

EI-MS (70 eV)

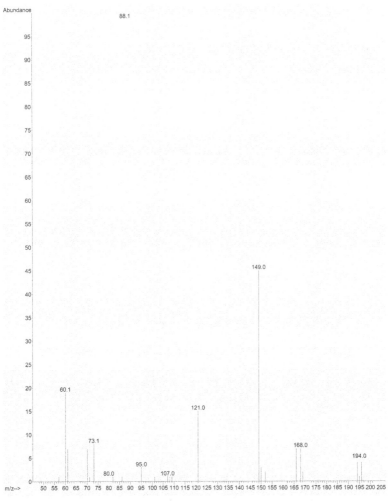

Figure 13.1 EI-MS of unknown compound **13**.

IR SPECTRUM (Thin film)

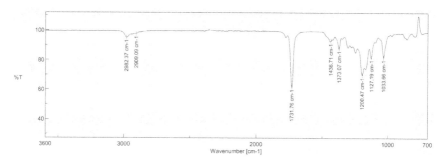

Figure 13.2   IR spectrum of unknown compound **13**.

¹H NMR SPECTRUM (500 MHz, CDCl₃)

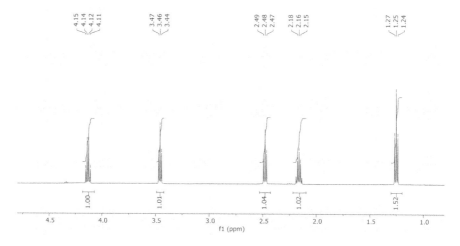

Figure 13.3   ¹H NMR spectrum of unknown compound **13**.

TABLE 13.1   ¹H NMR peaks of unknown compound **13**.

δ (ppm)	υ (Hz)	δ (ppm)	υ (Hz)	δ (ppm)	υ (Hz)
1.24	618.8	2.18	1088.3	3.46	1728.2
1.25	626.0	2.19	1095.0	3.47	1734.7
1.27	633.1	2.47	1233.0	4.11	2055.8
2.13	1067.7	2.48	1240.2	4.12	2062.8
2.15	1074.6	2.49	1247.4	4.14	2070.0
2.16	1081.3	3.44	1721.8	4.15	2077.2

¹³C{¹H} NMR SPECTRUM (125 MHz, CDCl₃)

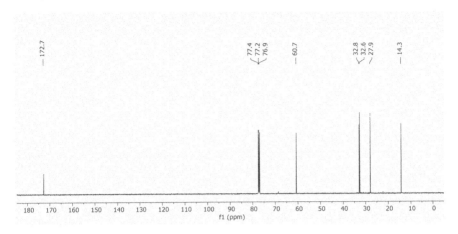

Figure 13.4 ¹³C{¹H} NMR spectrum of unknown compound **13**.

¹H-¹H COSY SPECTRUM

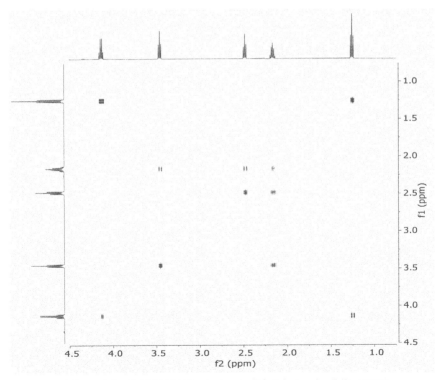

Figure 13.5 ¹H-¹H COSY spectrum of unknown compound **13**.

¹H-¹³C HSQC SPECTRUM

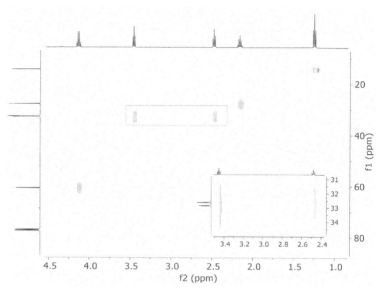

Figure 13.6   ¹H-¹³C HSQC spectrum of unknown compound **13**.

¹H-¹³C HMBC SPECTRUM

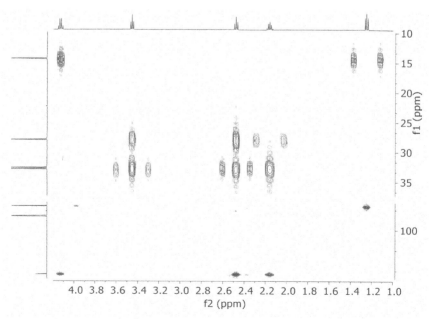

Figure 13.7   ¹H-¹³C HMBC spectrum of unknown compound **13**.

Problem 14

A by-product generated in the synthesis of hepialone, a sex pheromone of the moth *Hepialus Californicus B*, was fully characterized by the chemists who synthesized the target compound. Propose a structure for the above by-product given the following spectra.

EI-MS (70 eV)

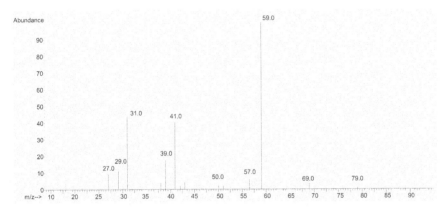

Figure 14.1   EI-MS of unknown compound **14**.

IR SPECTRUM (Thin film)

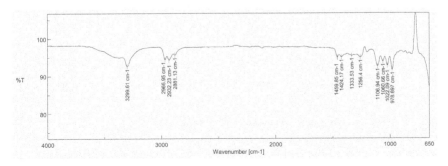

Figure 14.2   IR spectrum of unknown compound **14**.

¹H NMR SPECTRUM (500 MHz, CDCl₃)

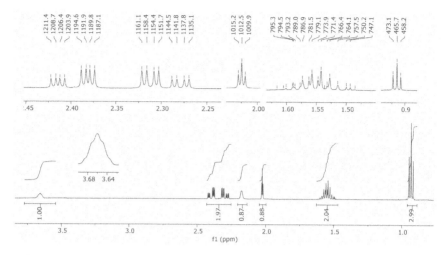

Figure 14.3   ¹H NMR spectrum of unknown compound **14**.

TABLE 14.1   ¹H NMR peaks of unknown compound **14**.

δ (ppm)	υ (Hz)	δ (ppm)	υ (Hz)	δ (ppm)	υ (Hz)
3.69	1844.0	2.31	1154.4	1.56	781.5
3.67	1836.8	2.30	1151.7	1.56	779.1
3.67	1834.4	2.29	1144.5	1.55	773.9
3.66	1830.5	2.28	1141.8	1.54	771.4
3.65	1823.8	2.27	1137.8	1.54	768.4
3.64	1818.2	2.27	1135.1	1.53	766.4
2.42	1211.4	2.18	1087.9	1.53	764.1
2.42	1208.7	2.03	1015.2	1.51	757.5
2.41	1206.4	2.02	1012.5	1.50	750.2
2.41	1203.9	2.02	1009.9	1.49	747.1
2.39	1194.6	1.60	800.7	0.95	473.1
2.38	1191.9	1.59	795.3	0.93	465.7
2.38	1189.8	1.59	794.5	0.92	458.2
2.37	1187.1	1.59	793.2		
2.32	1161.1	1.58	789.0		
2.32	1158.4	1.57	786.9		

¹³C{¹H} NMR AND DEPT-135 SPECTRA (125 MHz, CDCl₃)

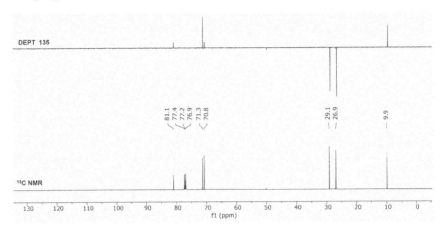

Figure 14.4 ¹³C{¹H} NMR and DEPT spectra of unknown compound **14**.

¹H-¹H COSY SPECTRUM

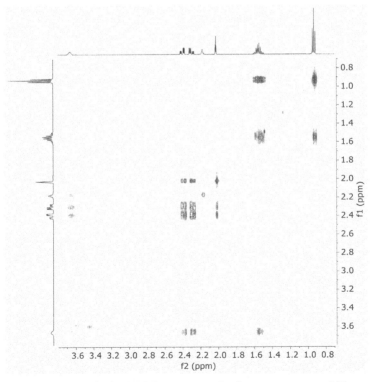

Figure 14.5 ¹H-¹H COSY spectrum of unknown compound **14**.

¹H-¹³C HSQC SPECTRUM

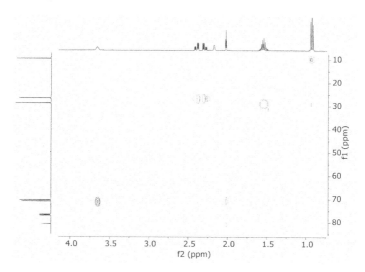

Figure 14.6   ¹H-¹³C HSQC spectrum of unknown compound **14**.

Problem 15

From which compound were the following spectra obtained?

EI-MS (70 eV)

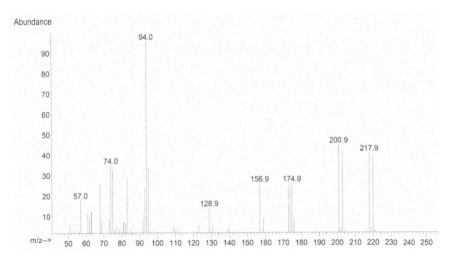

Figure 15.1   EI-MS of unknown compound **15**.

TABLE 15.1   Main peaks in EI-MS of unknown compound **15**.

m/z	Abundance (%)	m/z	Abundance (%)	m/z	Abundance (%)	m/z	Abundance (%)
51.0	6.0	75.0	33.0	95.0	34.0	174.9	26.0
57.0	18.0	77.0	4.0	108.9	5.0	175.9	9.0
61.0	11.0	78.9	4.0	123.0	5.0	200.9	45.0
62.0	10.0	80.9	7.0	128.9	14.0	202.9	42.0
63.0	12.0	82.0	7.0	130.9	5.0	217.9	41.0
68.0	26.0	83.0	29.0	156.9	26.0	218.9	4.0
69.0	8.0	92.0	9.0	158.9	9.0	219.9	39.0
73.0	9.0	93.0	24.0	172.9	25.0	220.9	3.0
74.0	35.0	94.0	100.0	173.9	24.0		

IR SPECTRUM (Thin film)

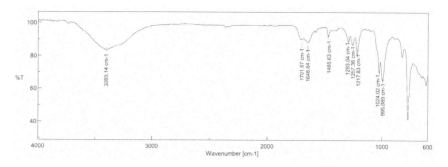

Figure 15.2   IR spectrum of unknown compound **15**.

¹H RMN SPECTRUM (500 MHz, CDCl₃)

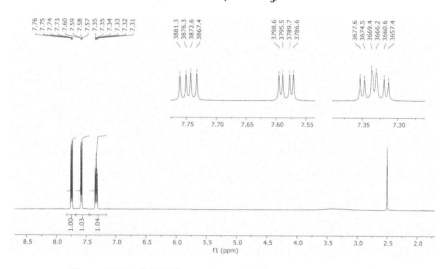

Figure 15.3   ¹H NMR spectrum of unknown compound **15**.

¹³C{¹H} NMR SPECTRUM (125 MHz, CDCl₃)

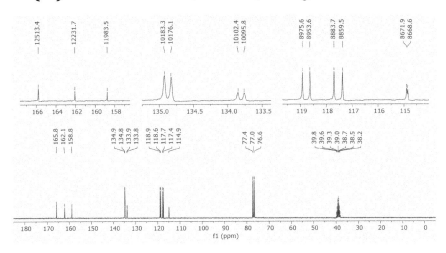

Figure 15.4 ¹³C{¹H} NMR spectrum of unknown compound **15**.

DEPT-135 NMR SPECTRUM (125 MHz, CDCl₃)

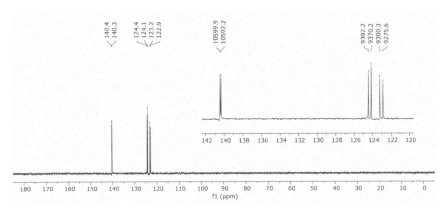

Figure 15.5 DEPT-135 NMR spectrum of unknown compound **15**.

¹H-¹³C HMBC SPECTRUM

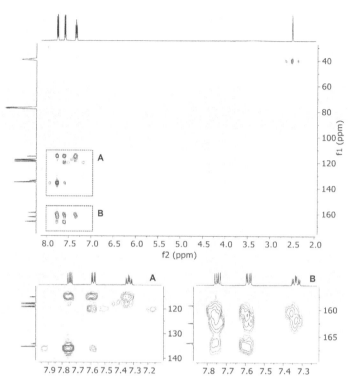

Figure 15.6   ¹H-¹³C HMBC spectrum of unknown compound **15**.

Problem 16

Construct a plausible structure for an antidepressant drug isolated as an orange powder given the following spectra.

EI-MS (70 eV)

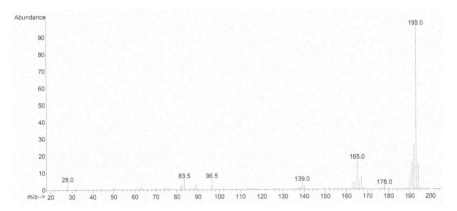

Figure 16.1 EI-MS of unknown compound **16**.

IR SPECTRUM (KBr plate)

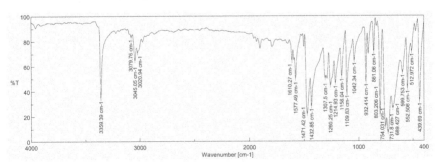

Figure 16.2 IR spectrum of unknown compound **16**.

¹H NMR SPECTRUM (500 MHz, DMSO-d₆)

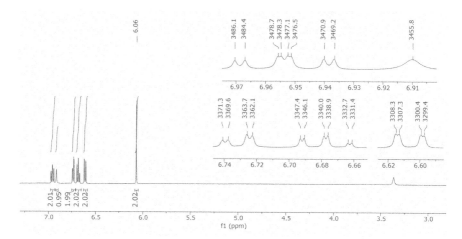

Figure 16.3   ¹H NMR spectrum of unknown compound **16**.

TABLE 16.1   ¹H NMR peaks of unknown compound **16**.

δ (ppm)	υ (Hz)	δ (ppm)	υ (Hz)	δ (ppm)	υ (Hz)
6.97	3486.1	6.91	3455.8	6.68	3338.9
6.97	3484.4	6.74	3371.3	6.66	3332.7
6.96	3478.7	6.74	3369.6	6.66	3331.4
6.95	3478.3	6.73	3363.7	6.61	3308.3
6.95	3477.1	6.72	3362.1	6.61	3307.3
6.95	3476.5	6.69	3347.4	6.60	3300.4
6.94	3470.9	6.69	3346.1	6.60	3299.4
6.94	3469.2	6.68	3340.0	6.06	3033.1

¹³C{¹H} NMR AND DEPT-135 SPECTRA (75.5 MHz, DMSO-d₆)

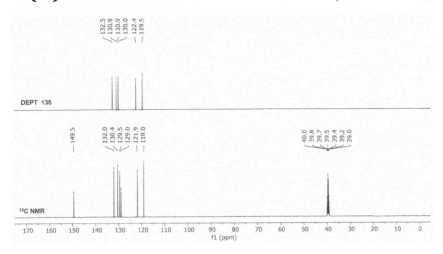

Figure 16.4 ¹³C{¹H} NMR and DEPT spectra of unknown compound **16**.

¹H-¹H COSY SPECTRUM

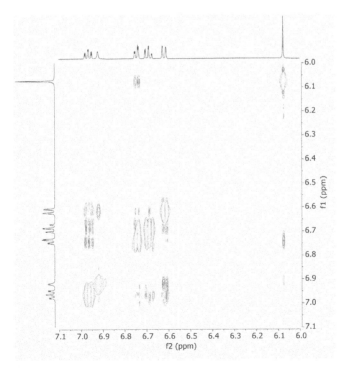

Figure 16.5 ¹H-¹H COSY spectrum of unknown compound **16**.

¹H-¹³C HSQC SPECTRUM

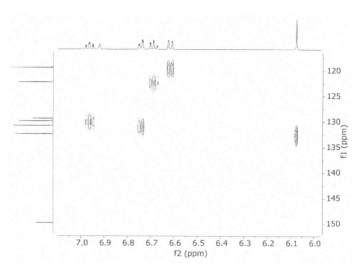

Figure 16.6   ¹H-¹³C HSQC spectrum of unknown compound **16**.

Problem 17

A Suzuki-Miyaura coupling reaction provided a colorless liquid as the main product. Propose a structure from the following spectra and assign as many signals as possible.

EI-MS (70 eV)

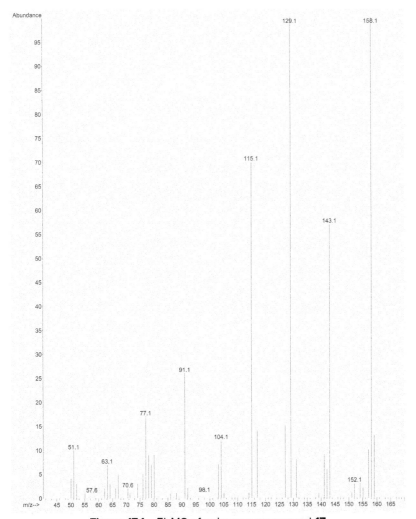

Figure 17.1 EI-MS of unknown compound **17**.

¹H RMN SPECTRUM (500 MHz, CDCl₃)

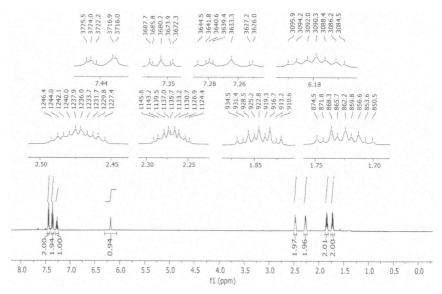

Figure 17.2 ¹H NMR spectrum of unknown compound **17**.

TABLE 17.1 ¹H NMR peaks of unknown compound **17**.

δ (ppm)	υ (Hz)	δ (ppm)	υ (Hz)	δ (ppm)	υ (Hz)
7.45	3725.5	6.18	3088.4	2.25	1126.9
7.45	3724.0	6.17	3086.2	2.25	1124.4
7.44	3722.2	6.17	3084.5	1.87	934.5
7.43	3716.9	2.49	1246.4	1.86	931.4
7.43	3716.0	2.49	1244.0	1.86	928.5
7.37	3687.7	2.48	1242.1	1.85	925.2
7.37	3685.8	2.48	1240.0	1.85	922.8
7.36	3680.2	2.48	1237.9	1.84	919.3
7.35	3673.9	2.47	1236.0	1.83	916.7
7.34	3672.3	2.47	1233.7	1.83	913.2
7.29	3644.5	2.46	1231.7	1.82	910.6
7.28	3641.8	2.46	1229.8	1.75	874.5
7.28	3640.6	2.45	1227.4	1.74	871.8
7.28	3639.4	2.29	1145.8	1.74	868.3
7.26	3633.3	2.29	1143.2	1.73	865.7
7.25	3626.0	2.28	1139.5	1.72	862.2
6.19	3095.9	2.27	1137.0	1.72	859.8
6.19	3094.2	2.27	1135.7	1.71	856.6
6.18	3092.0	2.27	1133.2	1.71	853.6
6.18	3090.3	2.26	1130.7	1.70	850.5

¹³C{¹H} NMR SPECTRUM (125 MHz, CDCl₃)

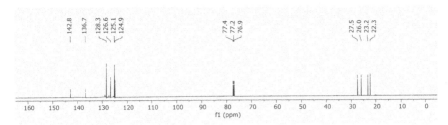

Figure 17.3 ¹³C{¹H} NMR spectrum of unknown compound **17**.

¹H-¹H COSY SPECTRUM

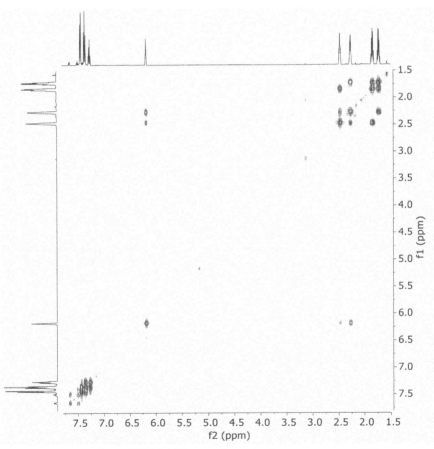

Figure 17.4 ¹H-¹H COSY spectrum of unknown compound **17**.

¹H-¹³C HSQC SPECTRUM

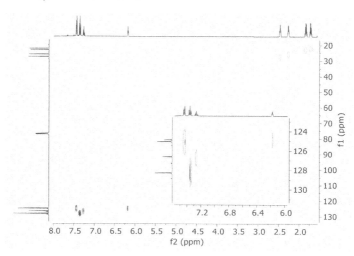

Figure 17.5   ¹H-¹³C HSQC spectrum of unknown compound **17**.

¹H-¹³C HMBC SPECTRUM

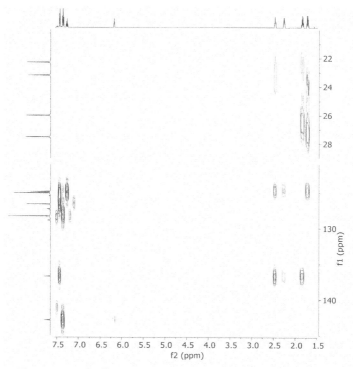

Figure 17.6   ¹H-¹³C HMBC spectrum of unknown compound **17**.

¹H-¹H NOESY SPECTRUM

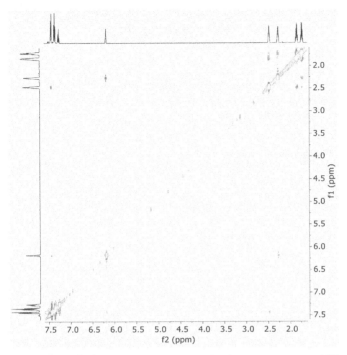

Figure 17.7   ¹H-¹H NOESY spectrum of unknown compound **17**.

Problem 18

Degradation of tryptophan by some bacteria provides a compound assayed as a neuroactive compound. Determine the structure of this potent scavenger of hydroxyl radicals from the spectra provided below.

EI-MS (70 eV)

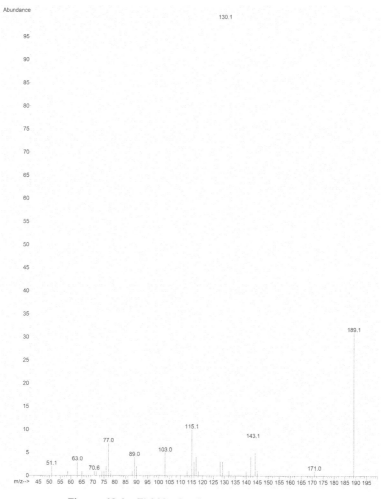

Figure 18.1 EI-MS of unknown compound **18**.

IR SPECTRUM (Thin film)

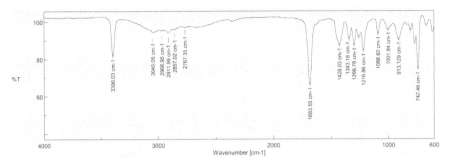

Figure 18.2 IR spectrum of unknown compound **18**.

¹H NMR SPECTRUM (500 MHz, dmso-d6)

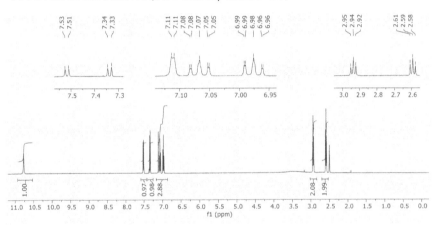

Figure 18.3 ¹H NMR spectrum of unknown compound **18**.

TABLE 18.1 ¹H NMR peaks of unknown compound **18**.

δ (ppm)	υ (Hz)	δ (ppm)	υ (Hz)	δ (ppm)	υ (Hz)
10.77	5388.0	7.05	3527.1	2.61	1303.7
7.52	3763.3	7.05	3526.2	2.59	1295.9
7.51	3755.4	6.99	3496.5	2.58	1288.4
7.34	3673.2	6.99	3495.6	2.51	1253.8
7.33	3665.0	6.98	3488.7	2.50	1252.1
7.11	3557.5	6.96	3481.6	2.50	1250.4
7.11	3555.4	6.96	3480.9	2.50	1248.4
7.08	3542.2	2.95	1476.2	2.49	1246.7
7.08	3541.1	2.94	1468.5		
7.07	3534.2	2.92	1461.1		

$^{13}C\{^1H\}$ NMR AND DEPT SPECTRA (125 MHz, dmso-d6)

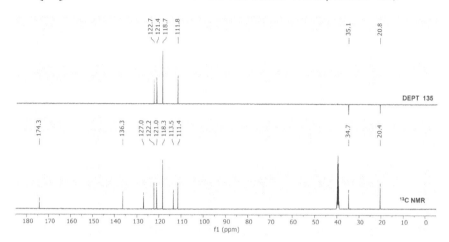

Figure 18.4 $^{13}C\{^1H\}$ NMR and DEPT spectra of unknown compound **18**.

^1H-^{13}C HSQC SPECTRUM

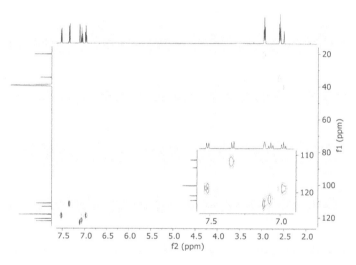

Figure 18.5 ^1H-^{13}C HSQC spectrum of unknown compound **18**.

¹H-¹³C HMBC SPECTRUM

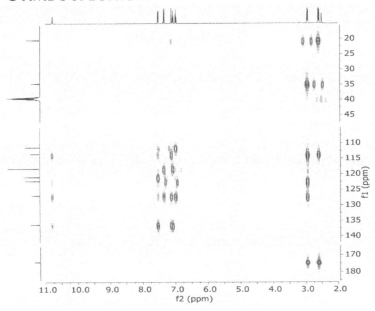

Figure 18.6 ¹H-¹³C HMBC spectrum of unknown compound **18**.

¹H-¹H NOESY SPECTRUM

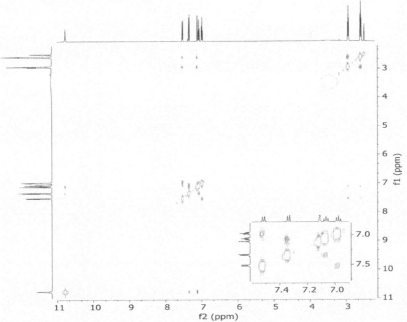

Figure 18.7 ¹H-¹H NOESY spectrum of unknown compound **18**.

Problem 19

From which compound, isolated as a white powder with pungent sour odor, were the following spectra obtained? Assign as many signals as possible.

EI-MS (70 eV)

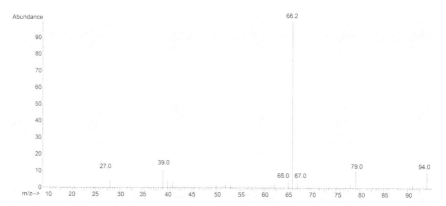

Figure 19.1   EI-MS of unknown compound **19**.

¹H NMR SPECTRUM (500 MHz, CDCl₃)

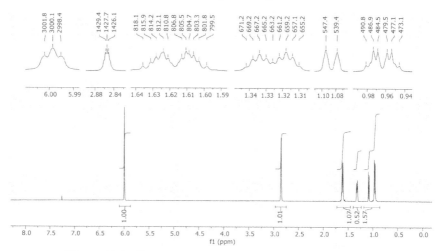

Figure 19.2   ¹H NMR spectrum of unknown compound **19**.

TABLE 19.1 ¹H NMR peaks of unknown compound **19**.

δ (ppm)	υ (Hz)	δ (ppm)	υ (Hz)	δ (ppm)	υ (Hz)
6.00	3001.8	1.61	806.8	1.32	659.2
6.00	3000.1	1.61	805.5	1.31	657.1
6.00	2998.4	1.61	804.7	1.09	547.4
2.86	1429.4	1.61	803.3	1.08	539.4
2.85	1427.7	1.60	801.8	0.98	490.8
2.85	1426.1	1.60	799.5	0.97	486.9
1.64	818.1	1.34	669.2	0.97	484.5
1.63	815.9	1.33	667.2	0.96	479.5
1.63	814.2	1.33	665.2	0.95	477.1
1.62	812.1	1.33	663.2	0.95	473.1
1.62	810.8	1.32	661.2		

¹³C{¹H} NMR SPECTRUM (125 MHz, CDCl₃)

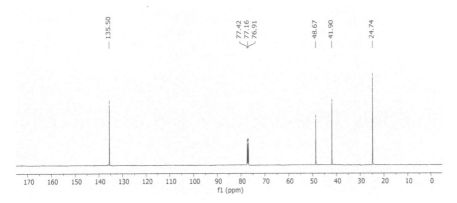

Figure 19.3 ¹³C{¹H} NMR spectrum of unknown compound **19**.

¹H-¹H COSY SPECTRUM

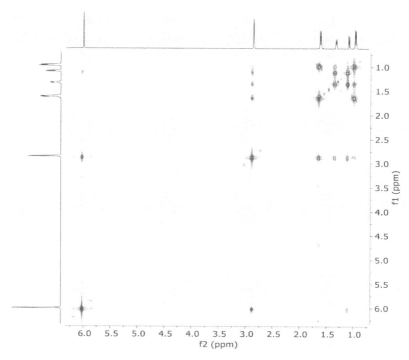

Figure 19.4   ¹H-¹H COSY spectrum of unknown compound **19**.

¹H-¹³C HSQC SPECTRUM

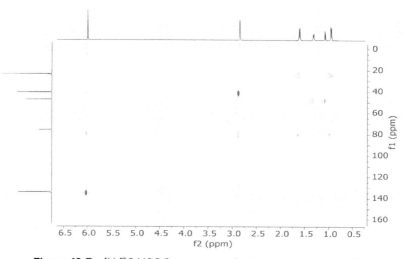

Figure 19.5   ¹H-¹³C HSQC spectrum of unknown compound **19**.

¹H-¹³C HMBC SPECTRUM

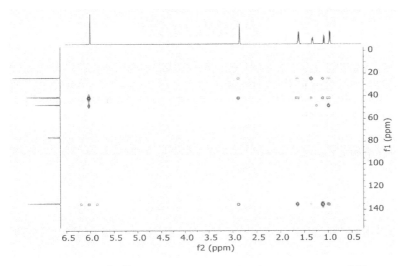

Figure 19.6   ¹H-¹³C HMBC spectrum of unknown compound **19**.

¹H-¹H NOESY SPECTRUM

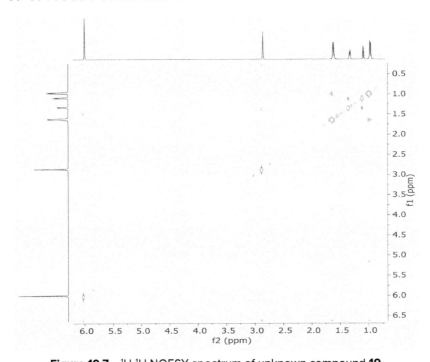

Figure 19.7   ¹H-¹H NOESY spectrum of unknown compound **19**.

Problem 20

A compound that participates in Arachidonic acid metabolism pathways was analyzed in order to determine its structure. Deduce its identity from the following spectra.

EI-MS (70 eV)

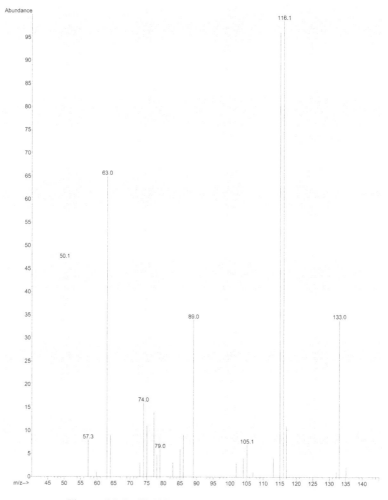

Figure 20.1 EI-MS of unknown compound **20**.

IR SPECTRUM (Thin film)

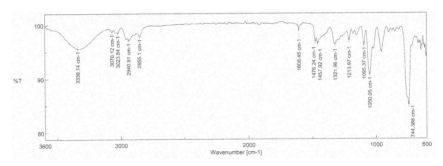

Figure 20.2 IR spectrum of unknown compound **20**.

¹H NMR SPECTRUM (300 MHz, CDCl₃)

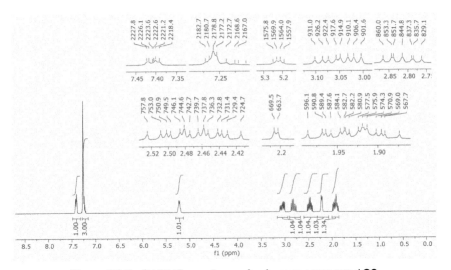

Figure 20.3 ¹H NMR spectrum of unknown compound **20**.

TABLE 20.1 ¹H NMR peaks of unknown compound **20**.

δ (ppm)	υ (Hz)	δ (ppm)	υ (Hz)	δ (ppm)	υ (Hz)
7.42	2227.8	3.03	910.1	2.44	731.4
7.42	2226.1	3.02	906.4	2.43	729.4
7.41	2223.5	3.00	901.6	2.41	724.7
7.41	2222.6	2.87	860.0	2.23	669.5
7.40	2221.2	2.84	853.3	2.21	663.6

Contd.

TABLE 20.1   Contd.

δ (ppm)	υ (Hz)	δ (ppm)	υ (Hz)	δ (ppm)	υ (Hz)
7.39	2218.4	2.84	851.7	1.99	596.1
7.27	2182.6	2.81	844.8	1.97	590.8
7.27	2180.7	2.79	837.3	1.96	589.4
7.26	2178.8	2.78	835.7	1.96	587.6
7.25	2177.2	2.76	829.1	1.95	584.2
7.24	2172.7	2.52	757.8	1.94	582.5
7.23	2168.6	2.51	753.0	1.94	580.9
5.25	1575.7	2.50	750.9	1.92	577.5
5.23	1569.9	2.50	749.5	1.92	575.9
5.21	1564.0	2.49	746.1	1.91	574.3
5.19	1558.0	2.48	744.6	1.90	570.9
3.10	931.0	2.47	742.7	1.90	569.0
3.09	926.1	2.46	739.8	1.89	567.7
3.07	922.4	2.46	737.7	1.87	562.4
3.06	917.6	2.45	736.3	2.44	731.4
3.05	914.9	2.44	732.8		

^{13}C{^{1}H} NMR AND DEPT SPECTRA (125 MHz, CDCl$_3$)

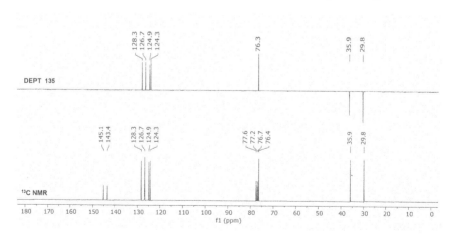

Figure 20.4 ^{13}C{^{1}H} NMR and DEPT spectra of unknown compound **20**.

¹H-¹H COSY SPECTRUM

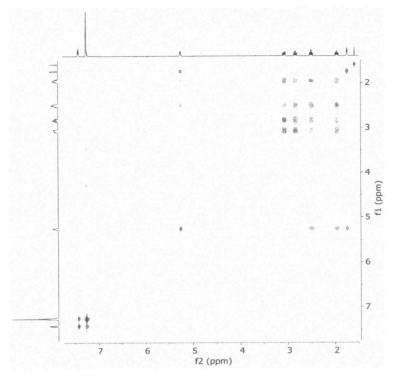

Figure 20.5 ¹H-¹H COSY spectrum of unknown compound **20**.

¹H-¹³C HSQC SPECTRUM

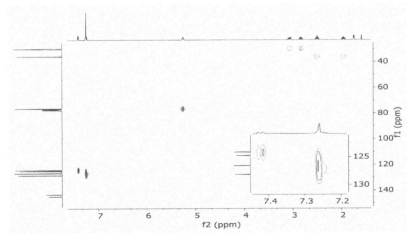

Figure 20.6 ¹H-¹³C HSQC spectrum of unknown compound **20**.

¹H-¹³C HMBC SPECTRUM

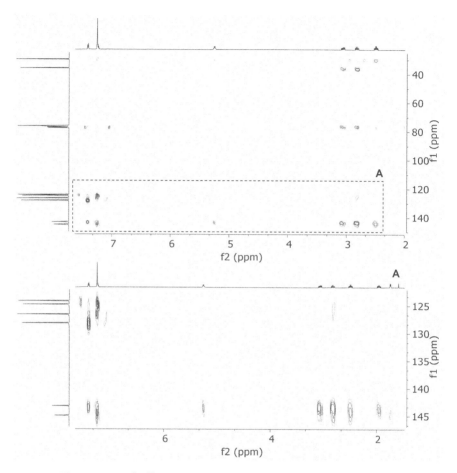

Figure 20.7   ¹H-¹³C HMBC spectrum of unknown compound **20**.

¹H-¹H NOESY SPECTRUM

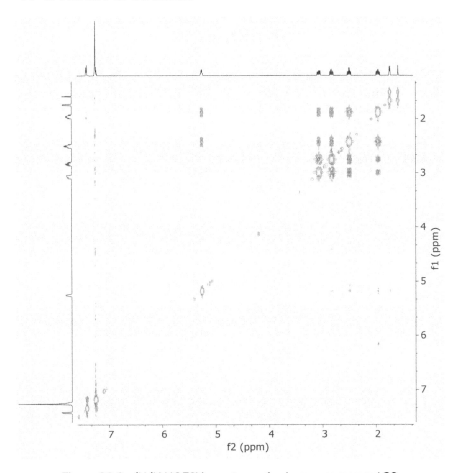

Figure 20.8 ¹H-¹H NOESY spectrum of unknown compound **20**.

Problem 21

The following spectra were recorded from a natural plant product found to inhibit tumor cells *in vitro*. Deduce its structure and assign as many signals as possible.

EI-MS (70 eV)

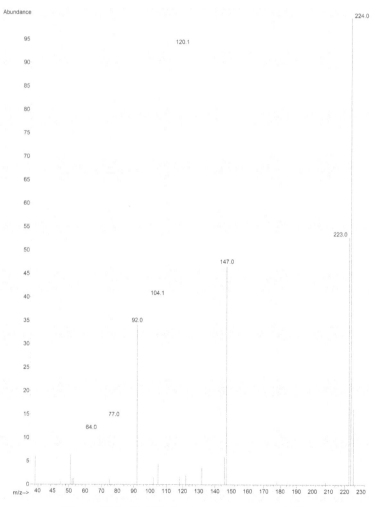

Figure 21.1 EI-MS of unknown compound **21**.

IR SPECTRUM (Thin film)

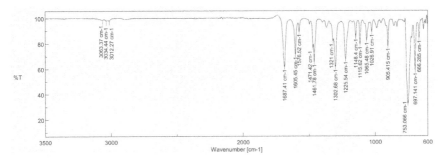

Figure 21.2   IR spectrum of unknown compound **21**.

¹H NMR SPECTRUM (500 MHz, CDCl₃)

TABLE 21.1   ¹H NMR peaks of unknown compound **21**.

δ (ppm)	υ (Hz)	δ (ppm)	υ (Hz)	δ (ppm)	υ (Hz)
7.96	3978.8	7.44	3722.5	7.06	3529.0
7.95	3977.0	7.43	3717.8	7.05	3525.6
7.94	3970.7	7.43	3716.5	7.05	3524.7
7.94	3969.0	7.41	3706.6	5.51	2754.8
7.53	3767.7	7.41	3705.2	5.50	2752.0
7.53	3766.0	7.41	3703.7	5.48	2741.5
7.52	3760.4	7.40	3700.4	5.48	2738.6
7.52	3759.4	7.39	3698.0	3.13	1564.3
7.51	3757.7	7.39	3695.3	3.10	1550.9
7.51	3754.5	7.38	3692.1	3.09	1547.3
7.50	3752.5	7.38	3690.7	3.07	1534.0
7.50	3750.5	7.38	3689.4	2.92	1462.1
7.49	3745.7	7.08	3539.9	2.92	1459.2
7.46	3732.1	7.07	3538.3	2.89	1445.2
7.46	3731.2	7.07	3537.2	2.88	1442.4
7.46	3729.3	7.06	3532.7		
7.45	3724.0	7.06	3530.0		

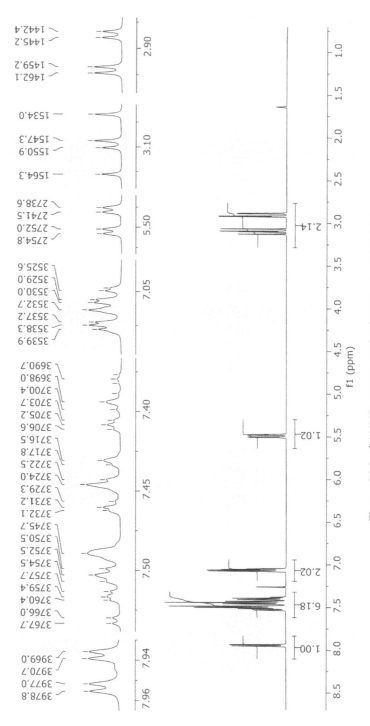

Figure 21.3   ¹H NMR spectrum of unknown compound **21**.

$^{13}C\{^1H\}$ NMR SPECTRUM (125 MHz, CDCl$_3$)

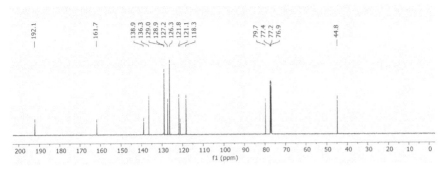

Figure 21.4 ^{13}C NMR spectrum of unknown compound **21**.

^{1}H-^{1}H COSY SPECTRUM

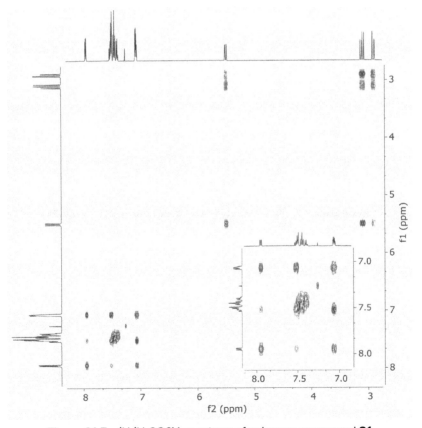

Figure 21.5 ^{1}H-^{1}H COSY spectrum of unknown compound **21**.

¹H-¹³C HSQC SPECTRUM

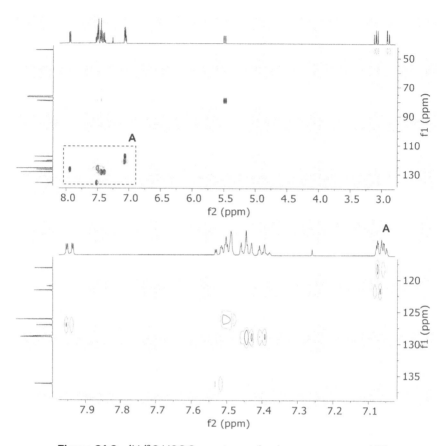

Figure 21.6   ¹H-¹³C HSQC spectrum of unknown compound **21**.

¹H-¹³C HMBC SPECTRUM

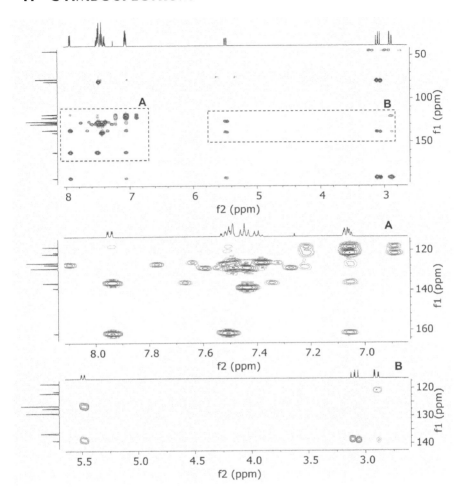

Figure 21.7 ¹H-¹³C HMBC spectrum of unknown compound **21**.

¹H-¹H NOESY SPECTRUM

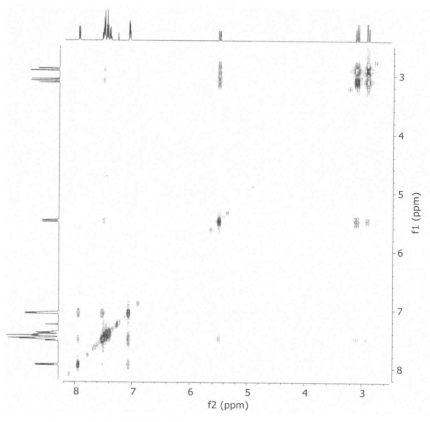

Figure 21.8   ¹H-¹H NOESY spectrum of unknown compound **21**.

Problem 22

The following spectra were recorded from an insect pheromone used for the control of bark beetles. Construct a plausible structure for this naturally occurring terpene with a pleasant odor.

EI-MS (70 eV)

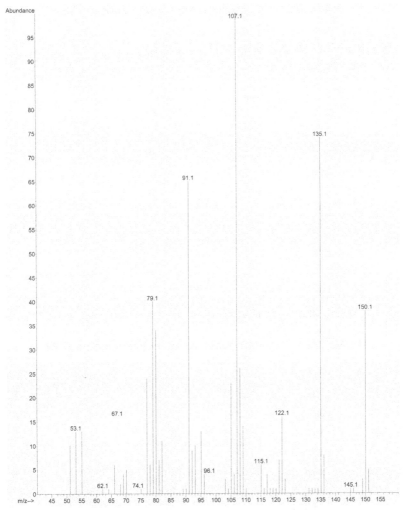

Figure 22.1   EI-MS of unknown compound **22**.

IR SPECTRUM (Thin film)

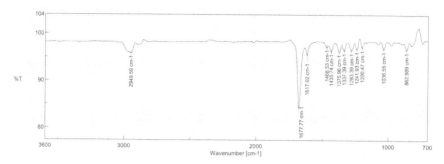

Figure 22.2   IR spectrum of unknown compound **22**.

¹H NMR SPECTRUM (500 MHz, CDCl₃)

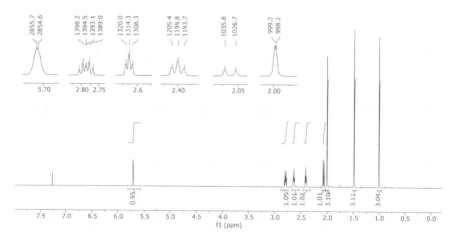

Figure 22.3   ¹H NMR spectrum of unknown compound **22**.

TABLE 22.1   ¹H NMR peaks of unknown compound **22**.

δ (ppm)	υ (Hz)	δ (ppm)	υ (Hz)	δ (ppm)	υ (Hz)
5.71	2855.7	2.77	1383.6	2.07	1035.8
5.71	2854.6	2.64	1320.0	2.05	1026.7
2.81	1403.6	2.63	1314.3	2.00	999.2
2.80	1398.2	2.62	1308.3	2.00	998.2
2.79	1394.5	2.41	1205.4	1.48	740.4
2.79	1393.1	2.40	1199.8	0.99	497.4
2.78	1389.0	2.39	1193.7		

¹³C NMR SPECTRUM (125 MHz, CDCl₃)

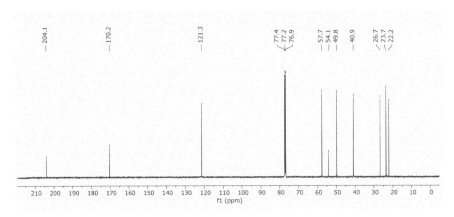

Figure 22.4 ¹³C NMR spectrum of unknown compound **22**.

¹H-¹H COSY SPECTRUM

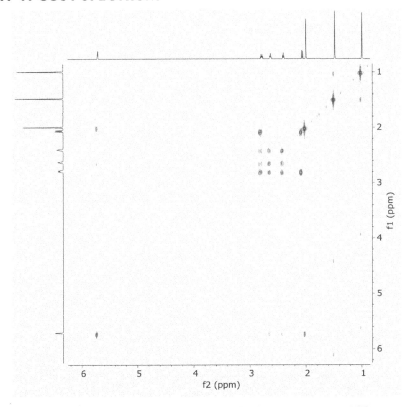

Figure 22.5 ¹H-¹H COSY spectrum of unknown compound **22**.

¹H-¹³C HSQC SPECTRUM

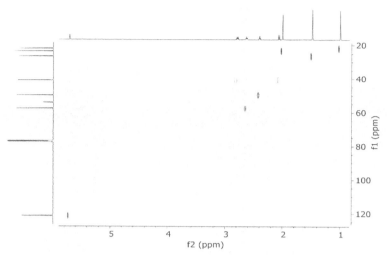

Figure 22.6 ¹H-¹³C HSQC spectrum of unknown compound **22**.

¹H-¹³C HMBC SPECTRUM

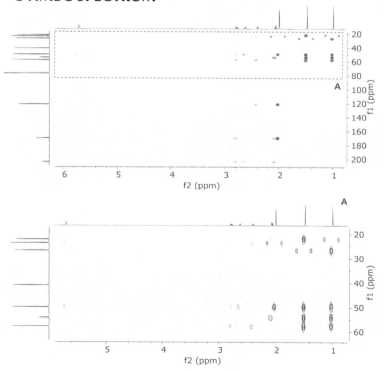

Figure 22.7 ¹H-¹³C HMBC spectrum of unknown compound **22**.

¹H-¹H NOESY SPECTRUM

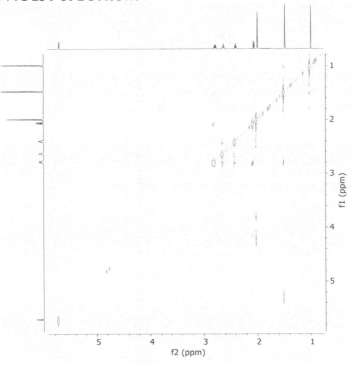

Figure 22.8   ¹H-¹H NOESY spectrum of unknown compound **22**.

Problem 23

Propose a structure from the spectra provided below. The compound was obtained as the main product of a palladium-catalyzed heteroannulation reaction.

HRMS: (M⁺) found 224.0837

¹H NMR SPECTRUM (300 MHz, CDCl₃)

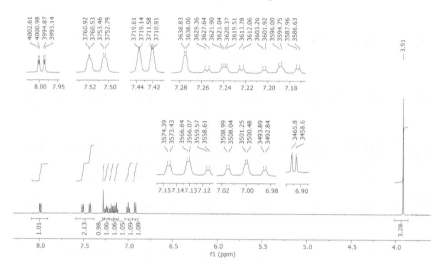

Figure 23.1 ¹H NMR spectrum of unknown compound **23**.

TABLE 23.1 ¹H NMR peaks of unknown compound **23**.

δ (ppm)	υ (Hz)	δ (ppm)	υ (Hz)	δ (ppm)	υ (Hz)
8.00	4002.6	7.25	3627.6	7.13	3566.8
8.00	4001.0	7.24	3621.9	7.13	3566.1
7.99	3994.9	7.24	3621.0	7.12	3559.6
7.98	3993.1	7.24	3620.4	7.12	3558.6
7.52	3760.9	7.24	3619.5	7.02	3509.0
7.52	3760.5	7.23	3613.8	7.01	3508.0
7.50	3753.5	7.22	3612.1	7.00	3501.2

Contd.

TABLE 23.1 Contd.

δ (ppm)	υ (Hz)	δ (ppm)	υ (Hz)	δ (ppm)	υ (Hz)
7.50	3752.8	7.20	3603.3	7.00	3500.5
7.44	3719.6	7.20	3601.9	6.99	3493.9
7.44	3719.1	7.19	3596.0	6.98	3492.8
7.42	3711.6	7.19	3594.8	6.93	3466.4
7.42	3710.9	7.17	3588.0	6.91	3458.2
7.28	3638.8	7.17	3586.6	3.91	1957.3
7.27	3638.1	7.15	3574.4		
7.26	3629.4	7.14	3573.4		

^{13}C NMR SPECTRUM (75.5 MHz, CDCl$_3$)

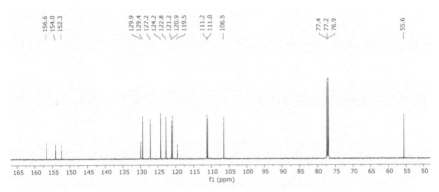

Figure 23.2 ^{13}C{^{1}H} NMR spectrum of unknown compound **23**.

^{1}H-^{1}H COSY SPECTRUM

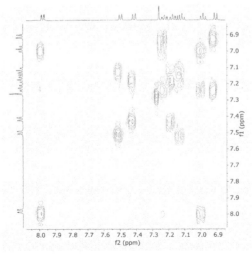

Figure 23.3 ^{1}H-^{1}H COSY spectrum of unknown compound **23**.

¹H-¹³C HSQC SPECTRUM

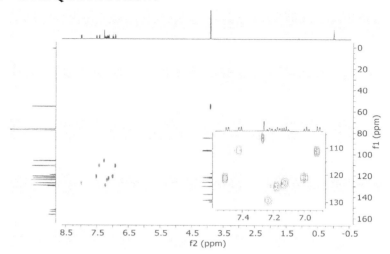

Figure 23.4   ¹H-¹³C HSQC spectrum of unknown compound **23**.

¹H-¹³C HMBC SPECTRUM

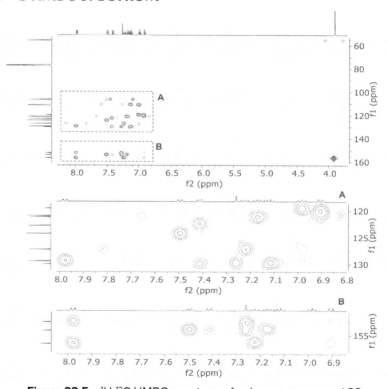

Figure 23.5   ¹H-¹³C HMBC spectrum of unknown compound **23**.

¹H-¹H NOESY SPECTRUM

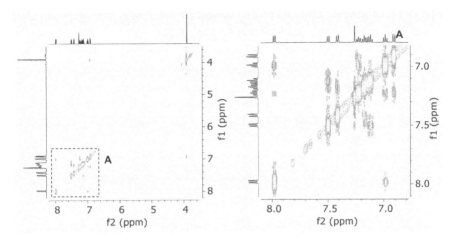

Figure 23.6   ¹H-¹H NOESY spectrum of unknown compound **23**.

Problem 24

A ligand used to prepare new ruthenium complexes catalytically active in organic media was analyzed in order to determine its structure. Deduce its identity from the following spectra.

EI-MS (70 eV)

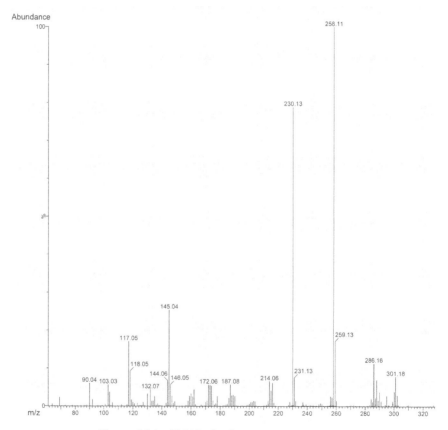

Figure 24.1 EI-MS of unknown compound **24**.

IR SPECTRUM (Thin film)

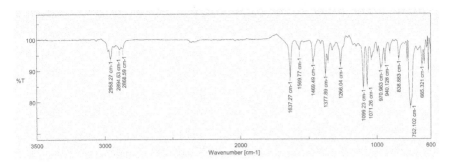

Figure 24.2 IR spectrum of unknown compound **24**.

¹H NMR (500 MHz, CDCl₃)

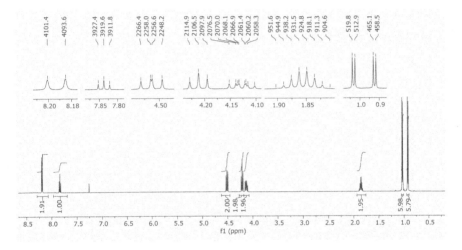

Figure 24.3 ¹H NMR spectrum of unknown compound **24**.

TABLE 24.1   ¹H NMR peaks of unknown compound **24**.

δ (ppm)	υ (Hz)	δ (ppm)	υ (Hz)	δ (ppm)	υ (Hz)
8.20	4101.4	4.19	2097.9	1.86	931.5
8.19	4093.6	4.15	2076.5	1.85	924.8
7.85	3927.4	4.14	2070.0	1.84	918.1
7.84	3919.6	4.14	2068.1	1.82	911.3
7.82	3911.8	4.13	2066.9	1.81	904.6
4.53	2266.4	4.12	2061.4	1.04	519.8
4.51	2258.0	4.12	2060.2	1.03	512.9
4.51	2256.6	4.10	2051.8	0.93	465.1
4.50	2248.2	1.90	951.6	0.92	458.5
4.23	2114.9	1.89	944.9		
4.21	2106.5	1.88	938.2		

¹³C NMR SPECTRUM (125 MHz, CDCl₃)

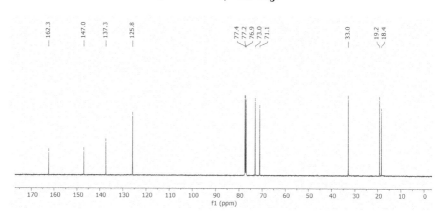

Figure 24.4   ¹³C{¹H} NMR spectrum of unknown compound **24**.

¹H-¹H COSY SPECTRUM

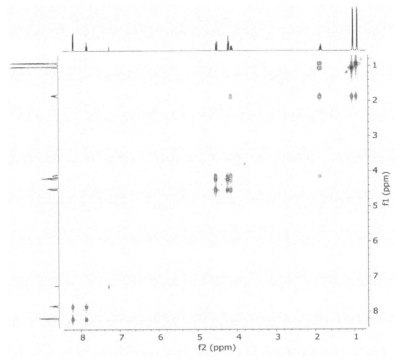

Figure 24.5 ¹H-¹H COSY spectrum of unknown compound **24**.

¹H-¹³C HSQC SPECTRUM

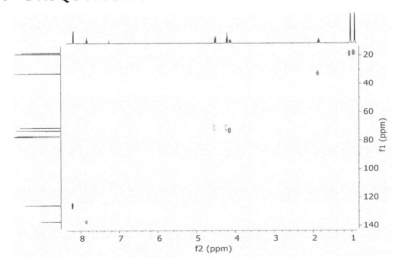

Figure 24.6 ¹H-¹³C HSQC spectrum of unknown compound **24**.

¹H-¹³C HMBC SPECTRUM

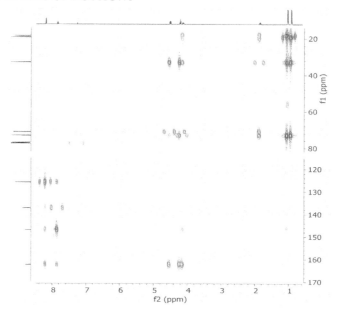

Figure 24.7   ¹H-¹³C HMBC spectrum of unknown compound **24**.

¹H-¹H NOESY SPECTRUM

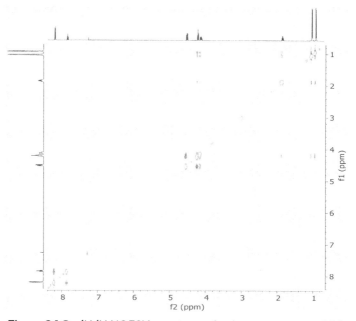

Figure 24.8   ¹H-¹H NOESY spectrum of unknown compound **24**.

Problem 25

Determine the structure of a moisture-sensitive substance whose spectra and spectral data are given below. The compound is known to be a potent alkylating agent.

EI-MS (70 eV)

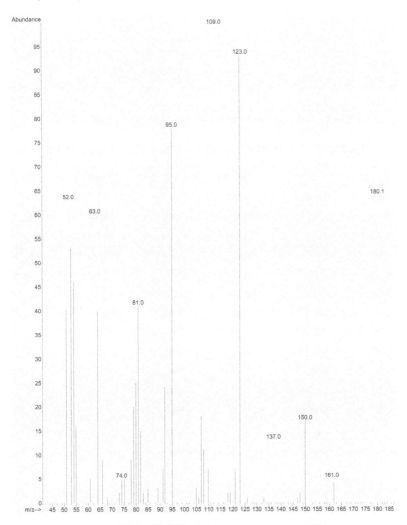

Figure 25.1 EI-MS of unknown compound **25**.

¹H NMR SPECTRUM (300 MHz, CDCl₃)

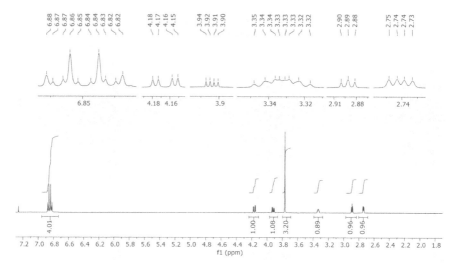

Figure 25.2   ¹H NMR spectrum of unknown compound **25**.

TABLE 25.1   ¹H NMR peaks of unknown compound **25**.

δ (ppm)	υ (Hz)	δ (ppm)	υ (Hz)	δ (ppm)	υ (Hz)
6.88	3440.0	4.16	2079.3	3.33	1664.7
6.87	3437.5	4.15	2076.0	3.32	1662.1
6.87	3433.5	3.94	1968.4	3.32	1658.8
6.86	3430.8	3.92	1962.8	2.90	1450.6
6.85	3427.7	3.91	1957.3	2.89	1445.8
6.84	3422.8	3.90	1951.8	2.88	1441.4
6.84	3419.7	3.77	1884.1	2.75	1374.5
6.83	3417.0	3.35	1674.5	2.74	1371.8
6.82	3413.0	3.34	1671.4	2.74	1369.5
6.82	3410.3	3.34	1668.6	2.73	1367.0
4.18	2090.4	3.33	1667.4		
4.17	2087.1	3.33	1665.9		

¹³C{¹H} NMR SPECTRA (125 MHz, CDCl₃)

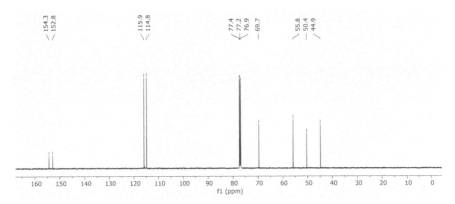

Figure 25.3 ¹³C{¹H} NMR spectrum of unknown compound **25**.

¹H-¹H COSY SPECTRUM

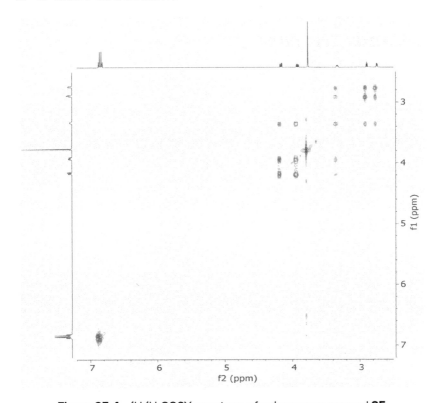

Figure 25.4 ¹H-¹H COSY spectrum of unknown compound **25**.

¹H-¹³C HSQC SPECTRUM

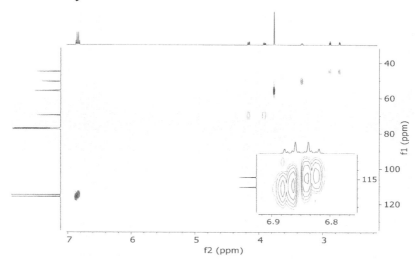

Figure 25.5   ¹H-¹³C HSQC spectrum of unknown compound **25**.

¹H-¹³C HMBC SPECTRUM

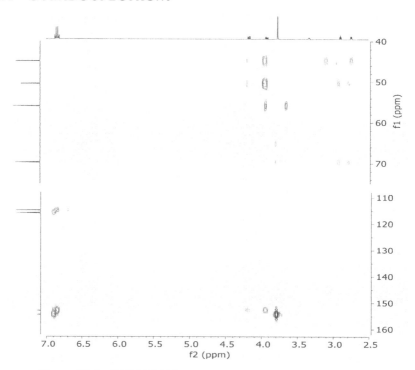

Figure 25.6   ¹H-¹³C HMBC spectrum of unknown compound **25**.

¹H-¹H NOESY SPECTRUM

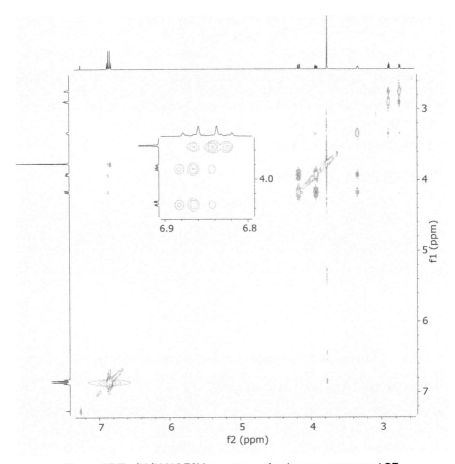

Figure 25.7 ¹H-¹H NOESY spectrum of unknown compound **25**.

Problem 26

An intramolecular palladium-catalyzed coupling produced a white powder fairly soluble in chloroform, from which the following spectra were recorded. What is the compound?

HRMS: (M⁺) found 235.0998

IR SPECTRUM (Thin film)

Figure 26.1   IR spectrum of unknown compound **26**.

¹H NMR SPECTRUM (500 MHz, CDCl₃)

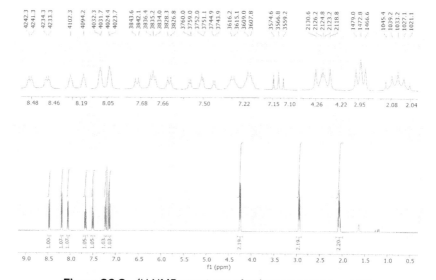

Figure 26.2   ¹H NMR spectrum of unknown compound **26**.

TABLE 26.1 ¹H NMR peaks of unknown compound **26.**

δ (ppm)	υ (Hz)	δ (ppm)	υ (Hz)	δ (ppm)	υ (Hz)
8.48	4242.3	7.65	3828.3	4.26	2130.6
8.48	4241.3	7.65	3826.8	4.25	2126.2
8.47	4234.3	7.52	3760.0	4.25	2124.8
8.46	4233.3	7.52	3759.0	4.25	2123.2
8.20	4102.3	7.50	3752.0	4.24	2118.8
8.19	4094.2	7.50	3751.1	2.96	1479.0
8.06	4032.3	7.49	3744.9	2.94	1472.8
8.06	4031.7	7.49	3743.9	2.93	1466.6
8.05	4024.4	7.23	3616.2	2.09	1045.4
8.05	4023.7	7.23	3615.1	2.08	1039.2
7.69	3843.6	7.22	3609.0	2.07	1033.2
7.68	3842.1	7.21	3607.8	2.05	1027.1
7.67	3836.4	7.15	3574.6	2.04	1021.1
7.67	3835.2	7.13	3566.8		
7.67	3834.0	7.12	3559.2		

¹³C{¹H} NMR SPECTRA (125 MHz, CDCl₃)

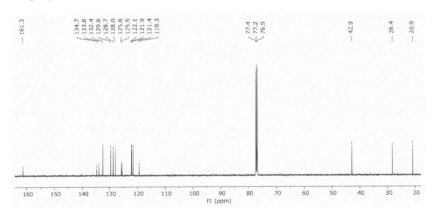

Figure 26.3 ¹³C{¹H} NMR spectrum of unknown compound **26.**

¹H-¹H COSY SPECTRUM

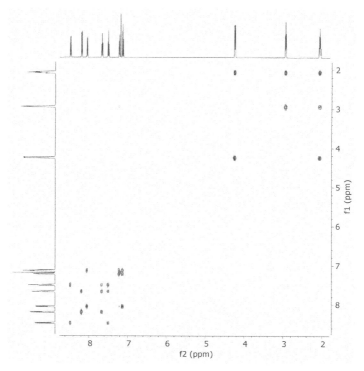

Figure 26.4   ¹H-¹H COSY spectrum of unknown compound **26**.

¹H-¹³C HSQC SPECTRUM

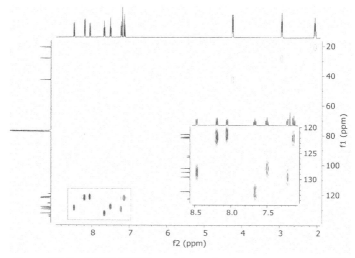

Figure 26.5   ¹H-¹³C HSQC spectrum of unknown compound **26**.

¹H-¹³C HMBC SPECTRUM

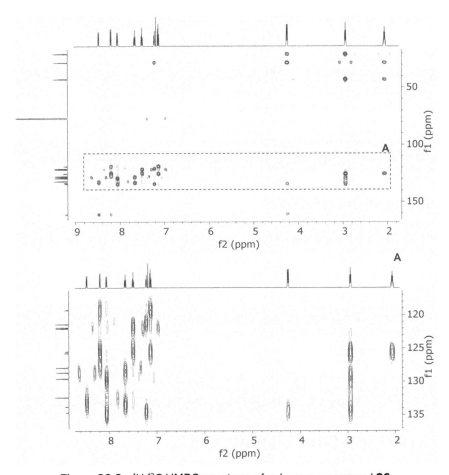

Figure 26.6 ¹H-¹³C HMBC spectrum of unknown compound **26**.

Problem 27

Thermal decomposition of aminoacids provided a complex mixture that was refluxed with several thiols in methanol under oxidative conditions. A fine white solid precipitated upon cooling. Deduce its identity from the following spectra and spectral data.

ESI-TOF HRMS: (MH⁺) found 346.1106

IR SPECTRUM (Thin film)

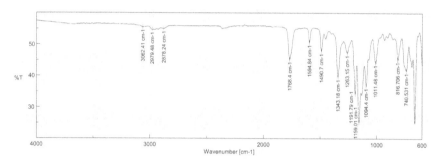

Figure 27.1 IR spectrum of unknown compound **27**.

¹H NMR SPECTRUM (500 MHz, CDCl₃)

TABLE 27.1 ¹H NMR peaks of unknown compound **27**.

δ (ppm)	υ (Hz)	δ (ppm)	υ (Hz)	δ (ppm)	υ (Hz)
7.74	3871.2	7.02	3510.6	2.14	1068.4
7.72	3863.0	7.02	3509.6	2.13	1064.2
7.32	3660.1	7.01	3507.1	2.13	1063.0
7.31	3658.0	4.42	2208.4	2.11	1056.0
7.31	3656.1	4.40	2202.7	2.10	1049.7
7.30	3650.5	4.40	2201.1	2.03	1017.5
7.30	3649.8	4.39	2195.4	2.02	1010.5
7.29	3644.0	3.50	1748.3	2.01	1005.1

Contd.

TABLE 27.1   Contd.

δ (ppm)	υ (Hz)	δ (ppm)	υ (Hz)	δ (ppm)	υ (Hz)
7.28	3642.1	3.49	1743.4	2.01	1003.2
7.28	3639.8	3.48	1740.9	2.00	998.0
7.25	3623.6	3.48	1738.8	1.99	996.1
7.23	3615.5	3.47	1735.9	1.99	994.8
7.18	3592.4	3.47	1733.8	1.98	990.8
7.17	3587.8	3.46	1731.3	1.96	982.3
7.17	3586.7	3.45	1726.4	1.95	974.9
7.17	3585.7	3.32	1659.6	1.80	899.0
7.16	3579.2	3.30	1652.4	1.79	893.4
7.14	3572.9	3.30	1650.1	1.78	892.1
7.14	3571.8	3.29	1645.1	1.77	886.7
7.14	3570.8	3.28	1642.8	1.76	881.2
7.04	3519.4	3.27	1635.5	1.75	874.5
7.03	3518.2	2.34	1172.6	1.74	869.0
7.03	3516.1	2.15	1076.0	1.72	862.1
7.02	3511.5	2.14	1071.8		

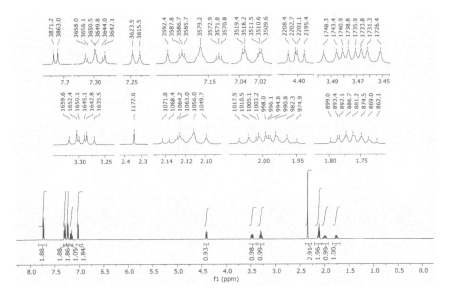

Figure 27.2   ¹H NMR spectrum of unknown compound **27**.

^{13}C{^{1}H} NMR SPECTRA (125 MHz, CDCl$_3$) AND DEPT 135

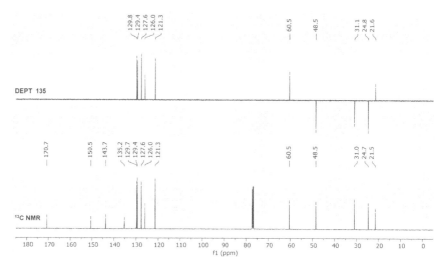

Figure 27.3 ^{13}C{^{1}H} NMR and DEPT spectra of unknown compound **27**.

^{1}H-^{1}H COSY SPECTRUM

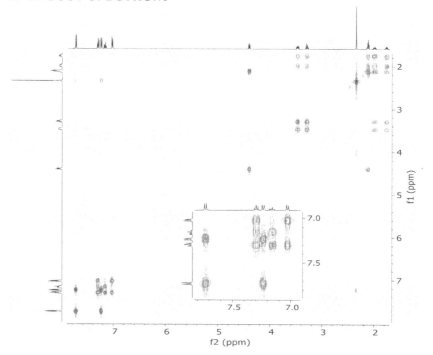

Figure 27.4 ^{1}H-^{1}H COSY spectrum of unknown compound **27**.

¹H-¹³C HSQC SPECTRUM

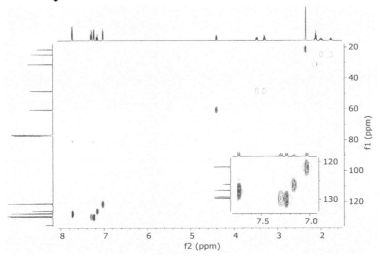

Figure 27.5   ¹H-¹³C HSQC spectrum of unknown compound **27**.

¹H-¹³C HMBC SPECTRUM

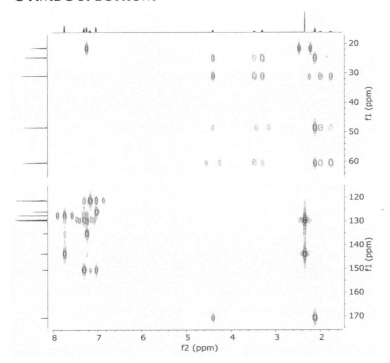

Figure 27.6   ¹H-¹³C HMBC spectrum of unknown compound **27**.

¹H-¹H NOESY SPECTRUM

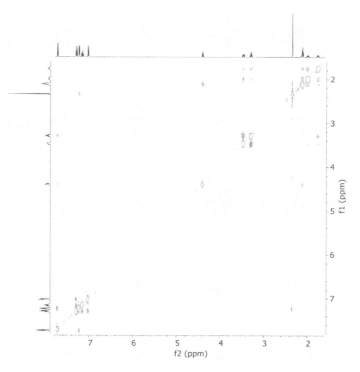

Figure 27.7   ¹H-¹H NOESY spectrum of unknown compound **27**.

SOLUTIONS TO PROBLEMS

Problem 1

A good starting point for structural determination is the molecular formula. Since no elemental analysis or high resolution mass spectrum is provided, we cannot determine it directly, but this is not necessarily an impediment to solve the structure. The few signals of the NMR spectra suggest that the molecule is small and, luckily, the mass spectrum provides valuable information. Indeed, the MS shows two peaks at m/z 122 and 124 with the same relative intensity that probably correspond to the molecular ion since they are the greatest m/z values in the spectrum, and the following ones respond to a loss of 15 mass units, usually the loss of a methyl group. The pattern consisting of two peaks of the same intensity differing by 2 mass units (M^+ and $M + 2$) indicates that it is a monobrominated molecule. The presence of this medium-heavy atom whose isotopes, ^{79}Br and ^{81}Br, are in a ratio of 51:49, would explain the molecular ions of the molecule. Moreover, the base peak at m/z 43 corresponds to the loss of bromine atom from the molecular ion and, therefore, supports our suspicion and confirms the small size of the carbon skeleton.

Although not really necessary, we could deduce the molecular formula of the compound by applying the rule of thirteen. Since $(9 \times 13) + 6$ is equal to 123, the formula of the corresponding hydrocarbon compound would be C_9H_{15}.[1] In order to include the bromine atom with an atomic mass of 79.90 uma, we should subtract six carbon and eight hydrogen atoms,[2] so the molecular formula is C_3H_7Br and the compound is undoubtedly a bromopropane.

For obvious reasons, the infrared spectrum shows no significant band outside the fingerprint region, apart from the stretching vibrations of the Csp^3-H single bonds at wavenumbers of 2900-3000 cm^{-1}. The stretching bands of C-Br bonds are difficult to identify because they appear inside the fingerprint area, specifically between 500 and 700 cm^{-1}.

The presence of only two signals in the ^{13}C NMR spectra gives us the last missing piece of the puzzle. The compound is symmetric so, it must be 2-bromopropane. Besides, both carbon signals are positive peaks in DEPT-135 experiment, thus confirming that there is no methylene carbon in the molecule. The presence of the electron withdrawing halogen shifts the signals downfield and they appear at δ 28.5 and 45.5 ppm. In addition,

---

[1] $123/13 = 9$ (R = 6) → $C_9H_{9+6} = C_9H_{15}$

[2] $6 \times 12\,(12C) + 8 \times 1(1H) = 80$

a six-proton doublet at 1.71 ppm (^{1}H NMR) indicates the presence of a methyl group attached to a carbon atom bound to only one proton. Although theoretically it should be a septet, the single-proton quintet at 4.29 ppm can only be assigned to the deshielded methine group attached to the bromine atom. Most likely, the outer lines are lost in noise. The coupling constant between methyl and methine protons ($^3J_{HH}$ = 6.6 Hz) is typical of the average coupling constants in freely rotating alkyl chains.

A full NMR signal assignment and a brief graphical explanation of the main fragmentations observed in the mass spectrum are shown in Figure 1.1.S.

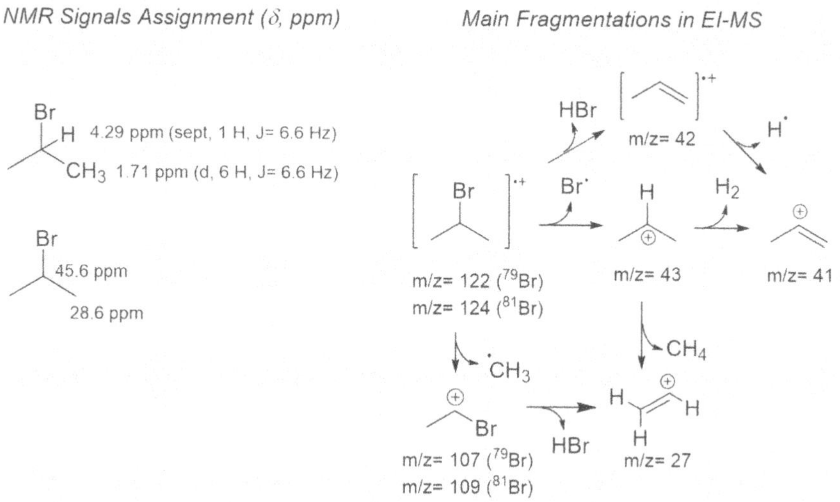

NMR Signals Assignment (δ, ppm)   *Main Fragmentations in EI-MS*

Figure 1.1.S Explanation of NMR and MS data of 2-bromopropane.

Problem 2

In this problem, a quick look at HRMS allows the easy calculation of the molecular formula by using an on-line tool such as ChemCalc (Patiny and Borel 2013). The formula obtained is $C_{14}H_{12}$. The high unsaturation index[3] (nine) suggests the presence of aromatic rings. Although the infrared spectrum is quite simple, it supports this suspicion by showing absorption bands at the frequency of the C_{sp^3}-H and C_{sp^2}-H single-bond vibrations (2833-3065 cm^{-1}) and C_{ar}-C_{ar} stretching bands at 1454 and 1485 cm^{-1}. Although clearly visible, no useful information for determining substitution patterns can be extracted from the region of overtones and combination bands.

The symmetry of the molecule is quite evident because there are only seven resonances in the ^{13}C NMR spectrum. Six out of these signals appear downfield, from 123.8 to 137.4 ppm. The magnetic anisotropy of an aromatic ring is likely to be the cause of such a high deshielding and, accordingly, the signals belong to aromatic carbons. DEPT-135 experiment shows only four resonances in the mentioned interval and, consequently, two aromatic carbons are quaternary. The seventh signal is a methylene carbon that resonates at 29.1 ppm, apparently shifted downfield by the aromatic rings. Taking into account this information as well as the molecular formula ($C_{14}H_{12}$), it can be easily deduced that two disubstituted benzenes are part of the structure. Those substituents can only be the above methylenes or the aromatic rings themselves. Both 9,10-dihydroanthracene and 9,10-dihydrophenantrene can be proposed as possible structures, but only the dihydrophenanthrene has the symmetry that would give rise to the number of signals observed in the ^{13}C NMR and ^{1}H NMR spectra (Fig. 2.1.S).

The ^{1}H NMR spectrum confirms the structure since it shows four signals; a two-proton singlet at 2.90 ppm, two multiplets between 7.24-7.36 ppm which integrate for three protons, and a one-proton doublet with an *ortho* aromatic coupling constant (7.6 Hz) at 7.78 ppm. The ^{1}H NMR of the 9,10-dihydroanthracene spectrum would be much simpler

---

[3] Unsaturation index, level of unsaturation or double-bond equivalent is the number of unsaturations present in a molecule. Instructions on how to calculate the degree of unsaturation can be found in "Problem-Solving Strategies" section in the Preface of the book.

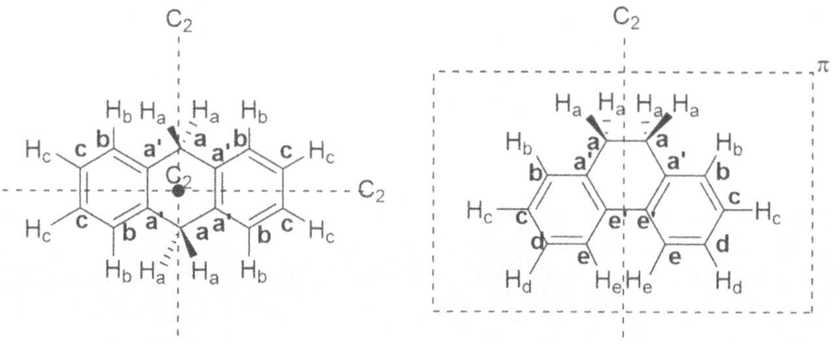

9,10-dihydroanthracene   9,10-dihydrophenantrene

Figure 2.1.S Symmetry and chemical equivalence of the nuclei of 9,10-dihydro-anthracene and 9,10-dihydrophenantrene.

than that because of the higher symmetry of the molecule. The signal at 2.90 ppm corresponds to a methylene that shows no coupling and it could be consistent with either of the structures because in none of them those protons have non-equivalent protons in the neighboring carbons. The chemical shift, however, suggests that the correct structure is the 9,10-dyhydrophenantrene because the bonding to the second aromatic ring would cause a greater downfield shift. Since there are no more experiments available to unequivocally assign NMR signals, we should use any of the available tools to estimate the chemical shift of the carbon and protons of the molecule, including tables of spectral data (Pretsch et al. 2010), on-line tools such as nmrpredictor (Binev et al. 2007, Banfi and Patiny 2008, Castillo et al. 2011, Aires-de-Sousa et al. 2002), or standalone software (ChemBioDraw 2012). The signals at δ_H 2.90 ppm and δ_C 29.1 ppm are directly attributed to the methylene unit and, after calculating the estimated chemical shifts, we could assign, without fear of being wrong, the doublet at 7.78 ppm to protons 4 and 5, and the signal at 137.4 ppm to carbons 4a and 4b, which are obviously the most deshielded nuclei in the molecule. The assignation of the rest of the aromatic proton and carbon signals is difficult without additional data (Fig. 2.2.S).

On the other hand, the loss of hydrogen, ethylene, and radicals that lead to the formation of stable benzylic cations explain the main fragmentation peaks observed in the mass spectrum (Fig. 2.2.S).

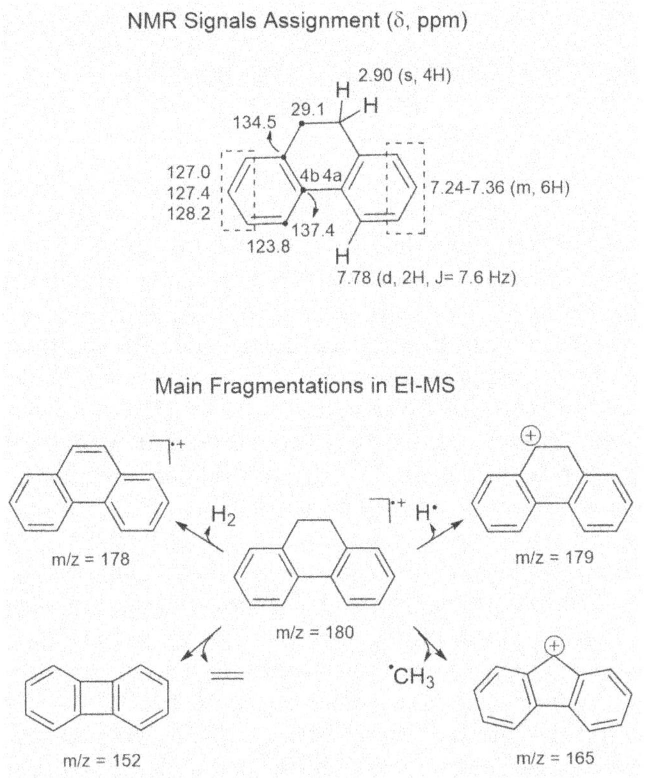

NMR Signals Assignment (δ, ppm)

2.90 (s, 4H)

134.5   29.1

127.0
127.4
128.2

4b 4a

7.24-7.36 (m, 6H)

137.4
123.8

7.78 (d, 2H, J= 7.6 Hz)

Main Fragmentations in EI-MS

m/z = 178

H₂

m/z = 180

H•

m/z = 179

m/z = 152

•CH₃

m/z = 165

Figure 2.2.S Assignment of the NMR signals and fragmentation pattern (MS) of 9,10-dihydrophenantrene.

Problem 3

Two mass spectra are available in this case. ESI mass spectra allows to identify the molecular ion and, subsequently, the molecular mass of the compound. Unfortunately, due to the fast fragmentation of molecular ions under high-energy EI conditions, in many EI-MS spectra, the molecular ion peak is too small to be detected. In this ESI-MS spectrum, we can easily observe the peaks corresponding to the protonated molecular ion[4] $[MH]^+$, m/z 76, and the $[M + Na]^+$ adduct, m/z 98. However, because of the absence of molecular fragmentation, little or no structural information is available from this technique - just the opposite of what EI-MS provides. The odd molecular mass of the molecular ion (m/z = 75) indicates that we presumably have a nitrogen compound. A $\Delta_{m/z}$ of 18 mass units between M^+ (ESI) and the highest m/z value in EI-MS (57) suggests the easy loss of water, an indicator of the presence of a hydroxyl group in the structure.

The infrared spectrum shows three wide bands in the range 3178-3357 cm^{-1}. Primary amines give rise to two bands, though they typically appear at 3500 and 3400 cm^{-1}. The low frequency of the bands, the presence of a third one, and the width of all of them could be due to the presence of the aforementioned OH group and the formation of intra-/intermolecular hydrogen bonds. The stronger the hydrogen bonds, the longer the O-H or N-H bonds, and the broader the corresponding absorption band. The IR spectrum also shows quite a broad band at 1597 cm^{-1} that would correspond to N-H bending.

From the rule of thirteen we can generate the following molecular formulas: C_5H_5, $C_4H_{13}N$, C_3H_9NO... The NMR spectra showing three non-equivalent carbon nuclei and nine protons, and the information extracted from the IR spectrum suggest C_3H_9NO as a probable molecular formula for the unknown liquid extracted from a shampoo, a compound with no unsaturations.

DEPT-135 evidences the presence of three methylene units and their relatively high chemical shifts (35.2, 39.4, and 59.9 ppm), which indicate attachment to heteroatoms that cause a downfield shift of these carbons. Obviously, the compound is 3-aminopropan-1-ol and the carbon resonating

---

[4] $[M + H]^+$ is also used to represent the protonated molecule, a term more accurately coined to define MH^+ (Naught et al. 1997).

at 60.1 ppm should be the one bound to the hydroxyl unit. The signal at 39.4 ppm is probably derived from the carbon attached to the amino group, and the third methylene unit, directly bound to two carbon atoms, is partially deshielded (δ_C: 35.2 ppm) because of the electrowithdrawing effect of the two heteroatoms.

The signals shown by the ^{1}H NMR spectrum confirm the structure. There are two-proton triplets at 3.71 (J = 5.6 Hz) and 2.87 ppm (J = 6.2 Hz), corresponding respectively to the oxygen-bound methylene and nitrogen-bound methylene units. The third methylene gives rise to an apparent quintuplet at 1.63 ppm. Although these protons are coupled to two pairs of non-equivalent protons, the similar coupling constants lead to signals akin to a quintuplet. On the other hand, there is a broad signal at 2.47 ppm that integrates to three protons. This signal must be due to the amine and alcohol protons, although these nuclei, in principle, are chemically non-equivalent. The position of the resonance of OH and NH protons is unpredictable because it depends on the extension of the hydrogen bonding, which is equally dependent on the temperature, concentration, and solvent employed, and therefore the chemical shift is not helpful in this case. However, the shape of the signal is a good hint. Protons attached to oxygen and nitrogen exchange easily with other X-H protons. The consequence is the widening of the signal and the vanishing of the coupling with neighboring protons if the exchange is fast. Such an effect is observed on this spectrum since no splitting is shown by methylene signals on behalf of alcohol or amine protons. Besides, if there are non-equivalent protons in the molecule that can change places, as in the case of 3-aminopropanol, the signals of these protons will be affected and become a single line if the exchange is very fast.

The peaks of the EI-MS confirm the structure by showing the loss of neutral molecules, such as water and formaldehyde and typical α-cleavages. These fragmentations are shown in Figure 3.1.S together with the assignment of the NMR signals.

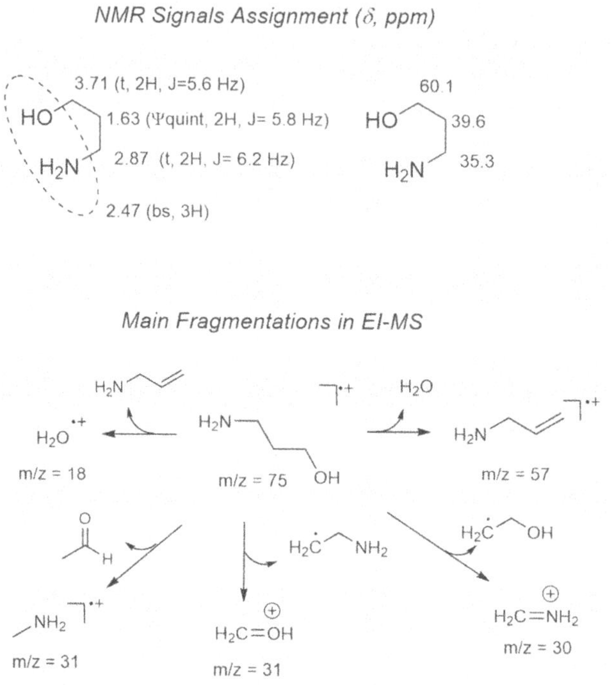

NMR Signals Assignment (δ, ppm)

3.71 (t, 2H, J=5.6 Hz)

1.63 (Ψquint, 2H, J= 5.8 Hz)

2.87 (t, 2H, J= 6.2 Hz)

2.47 (bs, 3H)

60.1

39.6

35.3

Main Fragmentations in EI-MS

m/z = 18

m/z = 75

m/z = 57

m/z = 31

m/z = 31

m/z = 30

Figure 3.1.S NMR signal assignment and main fragmentations in EI-MS of 3-aminopropanol.

Problem 4

The even molecular mass (M^+, m/z: 132) indicates that there is no nitrogen in the molecule (or at least not an odd number of nitrogen atoms). M^+-1 peak is the base peak of the spectrum, so there is a hydrogen in the molecule that is lost very easily. It is known, for instance, that the loss of the acetylenic hydrogen causes a strong M-1 peak in the mass spectrum of terminal alkynes.

Among the most straightforward molecular formulas calculated from the molecular mass using the rule of thirteen, C_9H_8O seems plausible, since ^1H NMR shows eight protons and there is a broad band in the infrared spectrum in the range of 3200-3500 cm^{-1}, which is probably the stretching band of an O-H single bond. Besides, the mass spectrum shows the loss of 17 (indicative of the presence of a hydroxyl group) at m/z 115. The double-bond equivalents or level of unsaturation of a compound whose molecular formula is C_9H_8O is seven, so the molecule is highly unsaturated.

The IR spectrums shows an absorption at 3291 cm^{-1} that overlaps with the aforementioned broad band. It could be the stretching of a C_{sp}-H bond, since mass spectrum shows a very strong M-1 peak. Moreover, a sharp absorption can be distinguished at approximately 2118 cm^{-1} where the stretching of the C≡C bond is normally observed. On the other hand, the high unsaturation index, the stretching of the aromatic C-C bands at 1451.2 and 1496.3 cm^{-1}, the overtone or combination bands with their characteristic pattern in the 2000-1650 cm^{-1} region, and the out-of-plane bending of the ring C-H bonds at 948, 739, and 699 cm^{-1} invite us to consider the presence of a an aromatic (aryl) ring in the molecule.

Since ^{13}C NMR shows 7 peaks, we know that some carbons in the molecule are chemically equivalent. Although we are aware that the area under the signal in a broadband proton decoupled ^{13}C NMR is not simply proportional to the number of carbons because the NOE is not equal for all the carbons[5], in view of the intensity of the signals, it could be suggested that the signals at 126.7 and 128.6 ppm are both the resonances of a pair of chemically equivalent carbons. The remaining two peaks at 140.0 and 128.4 ppm confirm the presence of an aromatic ring,

---

[5] For a discussion on the heteronuclear Overhausser effect responsible for this intensity enhancement, see "Basic One- and Two-Dimensional NMR Spectroscopy" (Friebolin 2010).

presumably, a monosubstituted benzene since only the resonance at 140.0 ppm corresponds to a quaternary carbon (DEPT-135). Furthermore, the above overtone or combination band pattern combined with the out-of-plane bending observed on the IR also evidence this point.

On the other hand, three carbon signals appear out of the aromatic region. One carbon resonates at 64.1 ppm, a chemical shift high enough to correspond to a CH carbon bound to a very electronegative atom such as oxygen. The two remaining resonances appear at 83.6 and 74.9 ppm, in the region of the alkyne carbons. The C≡C triple bond together with the benzene ring would match the unsaturation index of the molecule, and the mass and infrared spectra suggest the presence of an acetylenic hydrogen. However, at first sight, the DEPT-135 experiment contradicts this assumption since both carbons (83.6 and 74.9 ppm) show up in it. The DEPT experiment is the most widely used polarization transfer editing experiment in carbon-13 spectroscopy. The quaternary carbons do not show up in DEPT experiments because proton is excited first and the ^1H magnetization is transferred to the ^{13}C nucleus using the one-bond J coupling between the ^1H and the ^{13}C ($^1J_{CH}$ ~150 Hz). So, there is no signal if the carbon has no attached protons. Since the DEPT operates on the transfer of magnetization from ^1H to ^{13}C using $^1J_{CH}$, the experiment is set up for a particular J value. If there are an unusual J couplings in the molecule, a standard J setting may lead to errors in interpretation. Most J values are in the range of 125-170 Hz, so it is usually set to an average value of 145

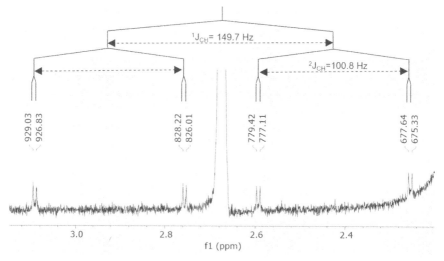

Figure 4.1.S   Expansion of the signal at 2.67 mm (^1H-NMR) of unknown compound **4**.

Hz suitable for most carbons. In terminal alkynes, $^1J_{CH}$ and $^2J_{CH}$ values are often unusually high, which frequently results in the vanishing of the terminal carbon in the DEPT 135 or the appearance of the quaternary digonal carbon due to a very large $^2J_{CH}$ (Jacobsen 2017, Claridge 2009).

The ^1H NMR allows us to clarify this matter since there is a one-proton doublet at 2.67 ppm whose chemical shift corresponds to an acetylenic proton, and the splitting pattern is consistent with a propargylic coupling (J = 2.2 Hz). The coupled proton appears further downfield (at 5.47 ppm), probably owing to the deshielding effect of a nearby oxygen and an aromatic ring. Interestingly, expansion of the acetylenic hydrogen signal displays the satellite bands due to those protons with an adjacent carbon 13. The signal is a doublet of doublet of doublets, and the calculated coupling constants are 149.7, 100.8, and 2.1 Hz, which correspond to $^1J_{CH}$, $^2J_{CH}$, and $^3J_{HH}$ couplings, respectively (Fig. 4.1.S).

Finally, a three-proton multiplet and a two-proton multiplet between 7.32 and 7.58 ppm confirm the presence of the monosubstituted benzene ring. Taking into account all of this information we can propose 1-phenyl-2-propyn-1-ol as the compound of Problem 4. As it can be seen in the ^1H NMR spectrum, the spin system is not first order, and therefore a detailed analysis of the signals and full assignment are difficult (Fig. 4.2.S).

NMR Signal Assignment (δ, ppm)

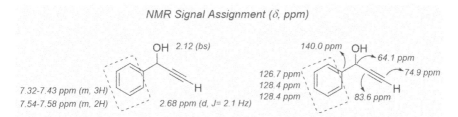

Figure 4.2.S NMR signal assignment of 1-phenyl-2-propyn-1-ol.

Analysis of the EI-MS spectrum confirms the proposed structure. In addition to the already mentioned loss of hydrogen atom and hydroxyl radical that provide peaks with odd m/z values (m/z 131 and 115 respectively), the loss of carbon monoxide and acetylene explain the main peaks of the spectra (Fig. 4.3.S).

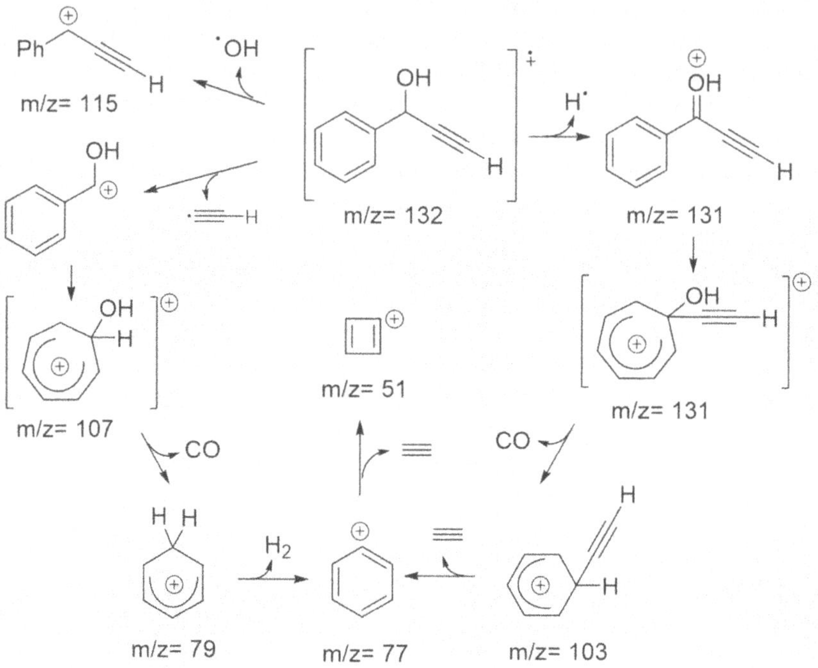

Figure 4.3.S Characteristic fragment ions in the EI mass spectrum of 1-phenylprop-2-yn-1-ol.

Problem 5

The molecular ion region of this compound has four peaks at m/z 182, 183, 184, and 185. One might assume that the highest m/z values correspond to the molecular ion (M^+) and the usual M + 1. However, we should not overlook that the peak at m/z = 182 is the base peak of the spectrum, and that we find another peak whose relative abundance is 33 separated by two units of m/z. This is the typical pattern of a chlorinated compound since the abundance of the chlorine-35 and chlorine-37 are approximately 76:24. The small peaks at m/z = 183 and 185 would be, therefore, the two M + 1 peaks (M + 1 and M + 3, to be precise). Although the peaks at m/z 182 and 184 seem to point out the presence of the chlorine atom in the molecule, there is no peak at m/z = M-35 (147) to confirm this hypothesis.

From the rule of thirteen, several molecular formulas can be proposed for this compound: $C_{14}H_{14}$, $C_{12}H_3Cl$, $C_{11}H_{15}Cl$, $C_{10}H_{11}ClO$, $C_9H_7ClO_2$, $C_7H_3ClO_2N_2$ ... The even molecular mass indicates no nitrogen, or at least no odd number of nitrogen atoms. ^{1}H-NMR shows that the number of protons is 3 or, rather, a multiple of 3. From ^{13}C-NMR, it is clear that there are seven non-equivalent carbons, so $C_7H_3ClO_2N_2$ is likely the molecular formula, from which we derive that the compound is highly unsaturated (d.b.e. = 7).

IR spectrum shows an absorption at 2233 cm^{-1} that should correspond to the C≡N stretch of a nitrile or a C≡C bond. On the other hand, the Csp^2-Csp^2 stretching band at 1560 as well as the Csp^2-H vibration frequencies (3095 and 3077 cm^{-1}) suggest the presence of an aromatic compound.

As mentioned before, the ^{13}C NMR spectrum shows 7 peaks, all of them between 110 and 150 ppm so the compound is likely aromatic. A glance at DEPT-135 experiment reveals that only the carbon nuclei at 129.2, 133.4, and 136.5 ppm are attached to proton. The carbon at 148.6 is particularly downfield, which is probably because it is directly attached to an electron-withdrawing atom or functional group. A chemical shift of approximately 150 ppm would make us think of a C_{arom}-O or an aromatic carbon attached to a strongly deshielding nitrogen group, such as nitro group. The last option is quite probable since the unsaturation index of the compound is 7, there are only seven carbons in the molecule, and apparently, there are two nitrogen and two oxygen atoms in the structure. If the aromatic ring is a trisubstituted benzene, the fourth quaternary carbon would be one of a nitrile group. In addition to the sharp absorption at 2233 cm^{-1} (IR) indicative of a cyano group, the presence of a nitro group is supported by the strong bands at 1560 and 1363 cm^{-1}, which are characteristic asymmetric and symmetric stretching bands of the $-NO_2$ unit.

The chemical shift of a nitrile carbon is quite insensitive to the chemical environment, and it usually falls between 113 and 120 ppm. Besides, the anisotropy of the C-N triple bond creates a shielding cone which has a strong effect on the carbon attached to it. The two quaternary carbons that resonate at 115.9 and 112.3 ppm would be those of the nitrile group and the quaternary aromatic carbon attached to it.

On the other hand, the ^{1}H NMR spectrum shows three one-proton signals in the aromatic region. One is a singlet at 8.18 ppm. Since our compound must be a trisubstituted benzene, the only possibility is that the spectrum is not resolved enough to show the smaller J_{meta} and J_{para} coupling constants so that this proton is not *ortho*-coupled. The other two signals are doublets with J = 8.4 Hz, which is a typical *ortho* coupling in benzenoid rings. The symmetry and intensity of these doublets is distorted, and the leaning (the inner lines are larger and the outer ones smaller) is evident, as it usually happens in second-order multiplets. The coupling constant is directly measurable but, as shown in Figure 5.1.S, line positions are no longer symmetrically related to chemical shift positions, and the chemical shift of the protons must be calculated in these cases. Some call this pattern "AB quartet", though it corresponds to the signals of two non-equivalent protons.

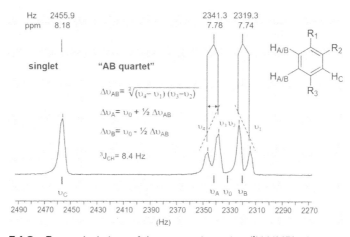

Figure 5.1.S Expanded view of the aromatic region (^{1}H NMR) of compound **5**.

Regarding the chemical shifts, we should take into account that nitro groups exhibit a pronounced electronic effect that causes *para* protons and *ortho* protons to be strongly downfield shifted (Williams and Fleming 2008). The presence of chlorine and nitrile subtituents has a much smaller deshielding effect even though they are also electron-withdrawing groups. Therefore, the nitro group is likely next to the most deshielded proton, that is, the one at 8.18 ppm. Although it is difficult to draw conclusions in polysubstituted benzenes, an analogous chemical shift would be expected for another proton *ortho* to the nitro group. So, we can think that the carbon attached to the nitro group is next to the CH unit, which shows no ^{1}H-^{1}H *ortho*-coupling, and to another quaternary aromatic carbon. The effect of nitro groups on carbon chemical shifts is not straightforward, and while *para* carbon is shifted downfield, the *ortho* carbon is usually shifted upfield. A remarkably low chemical shift value, lower than the observed δ_C 112.3 ppm, would be expected if a carbon suffered the combined shielding effect of nitrile group and nitro groups on the neighboring carbons. So, probably nitro group and chlorine are next to each other. The lack of any other spectra leads us to perform an estimation of the NMR chemical shifts in order to choose among possible regioisomers and assign as many signals as possible. Although the values shown in Figure 5.2.S may vary slightly depending on the tool used to estimate the chemical shifts, the trend is maintained, thus concluding that compound **5** is 4-chloro-3-nitrobenzonitrile.

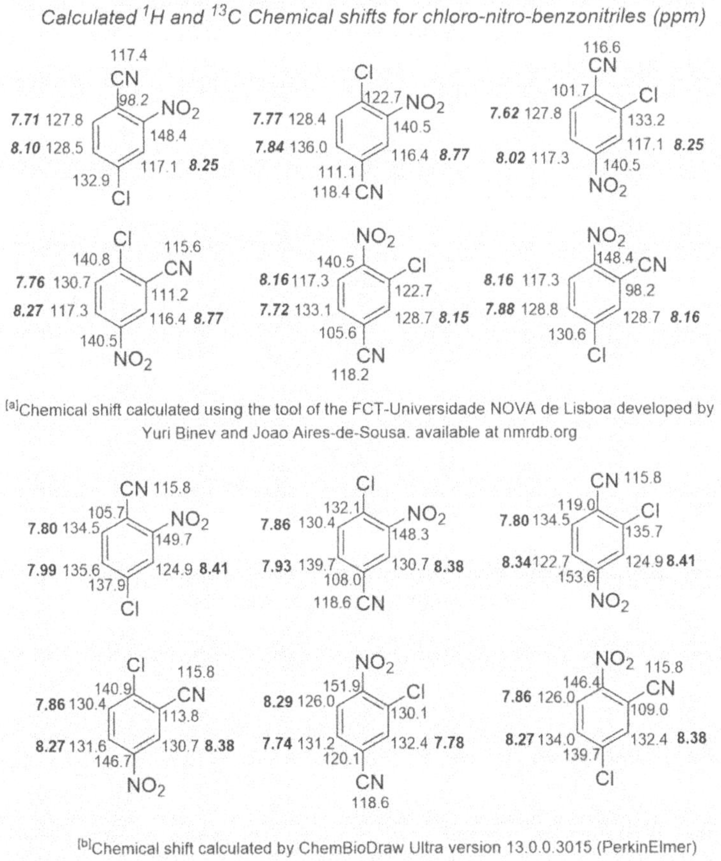

Calculated ^{1}H and ^{13}C Chemical shifts for chloro-nitro-benzonitriles (ppm)

[a]Chemical shift calculated using the tool of the FCT-Universidade NOVA de Lisboa developed by Yuri Binev and Joao Aires-de-Sousa. available at nmrdb.org

[b]Chemical shift calculated by ChemBioDraw Ultra version 13.0.0.3015 (PerkinElmer)

Figure 5.2.S Estimated chemical shifts for cloronitrobenzonitrile isomers.

Signal assignment based on both the calculated values and the data obtained from the ^1H and ^{13}C NMR spectra is shown in Figure 5.3.S.

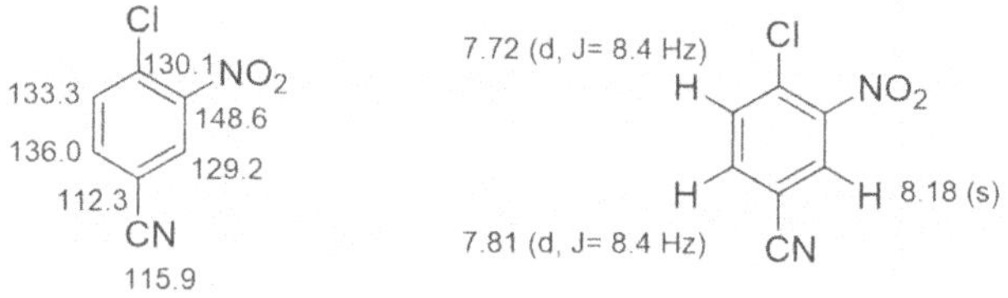

NMR Signal Assignment (δ, ppm)

Figure 5.3.S   Assignment of the NMR signals of 4-chloro-3-nitrobenzonitrile.

In order to explain the main peaks observed in the mass spectrum, the loss of NO_2 and NO radicals, and typical fragmentations of aromatic nitro compounds should be considered. Loss of neutral molecules such as CO and HCl provide other abundant peaks, including the base peak at m/z 100 (Fig. 5.4.S).

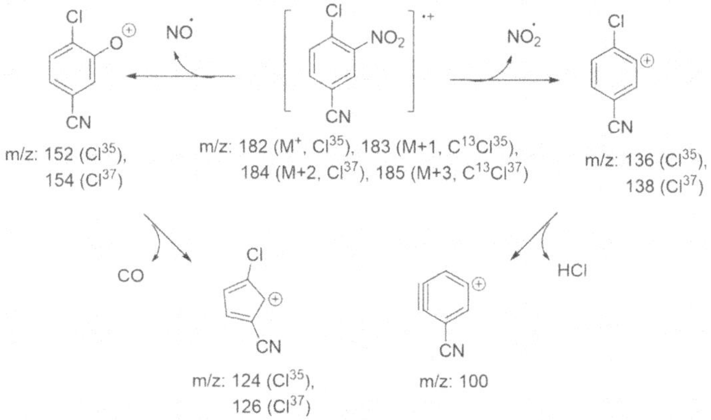

Figure 5.4.S   Main fragmentations in the EI-MS spectrum of 4-chloro-3-nitrobenzonitrile.

Problem 6

From the m/z value of the molecular ion of unknown compound **6** (m/z 318), quite a few molecular formulas can be deduced using the rule of thirteen (e.g., $C_{24}H_{30}$, $C_{23}H_{26}O$, $C_{22}H_{22}O_2$, $C_{22}H_{26}N_2$, $C_{20}H_{14}O_2S$, etc.). In this case, ^1H NMR and ^{13}C NMR are quite simple for a molecule whose mass is 318 mass units, so the symmetry of the molecule is obviously high. The protons of the compound are a multiple of 11 and the number of non-equivalent carbons is 9. Besides, the chemical shifts suggest that there are heteroatoms in the molecule. So, we can start working out which molecular formula fits the data or, we can just start "putting the pieces together" and think about what is missed later.

Apart from the absorptions around 3000 cm^{-1} (Csp^3-H and Csp^2-H stretching bands) and the weak bands at 1512 and 1591 cm^{-1} ($\nu_{Carom-Carom}$), there is no characteristic absorption out of the fingerprint region of the IR spectrum.

The evidence of the presence of an aromatic ring appears in ^{13}C NMR. There are 6 signals between 110 and 149 ppm, three of them corresponding to tertiary aromatic carbon. Two of the quaternary aromatic carbons are highly deshielded and resonate at 148.6 and 149.0 ppm, thus suggesting direct attachment to oxygen. The presence of oxygen atoms could also explain the high chemical shift of the aliphatic carbons at 55.9, 55.8, and 71.8 ppm. The latter is a strongly downshifted methylene (DEPT-135), probably deshielded by both a heteroatom and an aromatic ring.

The ^1H NMR shows two lines at 3.88 ppm, one singlet at 4.47 ppm, and a complex multiplet between 6.82 and 6.92 ppm, with the integration being $3:1:1.5$, respectively. Taking into account the signals of the ^{13}C NMR spectrum, we can say that there are three non-equivalent aromatic protons, a methylene in the molecule, and the two lines at 3.88 ppm correspond to the singlets of two methyl groups. The methylene protons resonate at 4.47 ppm and the two methyl groups at 3.88 ppm. These resonances are certainly the signals of two non-equivalent OMe groups attached to the aromatic ring and a strongly deshielded benzylic methylene due to the presence of an adjacent heteroatom. These considerations are also consistent with the two highly downshifted quaternary aromatic carbons (δ_C: 148.6 and 149.0 ppm) and the methylene at 72 ppm.

Regarding the signals of the three aromatic protons, it obvious that these nuclei are part of a second-order spin system. Since it is an aromatic system and the chemical shifts are very similar, an ABC spin system is expected. However, although the line intensities are no integral ratios and multiplets lean toward each other, the observed lines come close to an AXY system. Taking into account that the *ortho* and *meta* coupling constants involved are widely different in magnitude and that J_{para} is small (close to 0 Hz), this group of signals can be defined as consisting of a *meta* doublet, an *ortho* doublet, and a doublet of doublets (*ortho*, *meta*). In other words, we are dealing with a 1,2,4-trisubstituted benzene ring. In Figure 6.1.S-A, an attempt of tree diagram is shown but, as we know, a 1st order interpretation is imprecise and the system should be analyzed as an ABC.

Putting together the pieces we already know, that is, a trisubtituted aromatic ring, two OMe groups, and a probable benzyl methylene attached to a neigboring oxygen, one would realize that just doubling or making this dimethoxylated benzyl fragment twice as large, we get a compound of 318 mass units (Fig. 6.1.S-B).

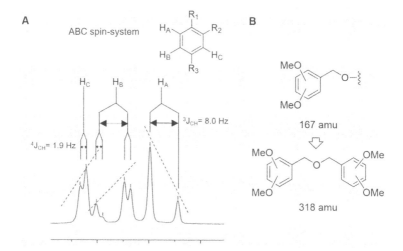

Figure 6.1.S   Aromatic region of ¹H NMR spectrum of unknown compound **6**.

In order to propose a plausible aromatic substitution pattern we have to consider first that all the carbons and protons are chemically not equivalent. The second point to be considered is the substitution of the ring (1, 2, 4), deduced from the study of the coupling constants between aromatic protons. Lastly, the chemical shifts of the nuclei need to be taken into account. As shown in Figure 6.2.S, the presence of an OMe group *ortho* to the benzylic methylene would downshift its ¹H-NMR signal, owing to a through-space deshielding by the methoxy oxygen. Besides, the chemical shift values of the quaternary carbons are crucial for a correct positioning of the OMe groups. These groups are resonance donors and, therefore, the nuclei in the *ortho* or *para* positions are upshifted because of a larger electron density. The chemical shift of the C_{arom}-O should have been higher if the two OMe groups were not next to each other. Moreover, the presence of one of them next to the benzyl methylene would cause the shielding of the C_{arom}-C, so that the chemical shift of this quaternary carbon should be lower. This is why we can conclude that the unknown molecule is veratryl ether (3,4-dimethoxybenzyl ether, Fig. 6.2.S).

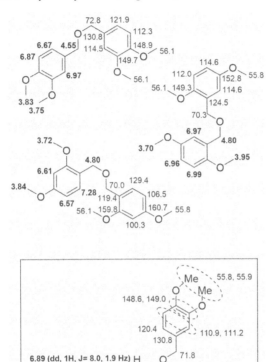

Figure 6.2.S Estimation of ^{1}H and ^{13}C chemical shifts (ppm) of dimethoxybenzyl ethers and assignment of the NMR signals of veratryl ether.

The mass spectrum also confirms the structure, and the base peak, m/z = 152, is the result of a McLafferty-type rearrangement that occurs with the loss of 3,4-dimethoxybenzaldehyde. The rest of the peaks can be reasonably explained by other related rearrangements based on hydrogen transfer and β-cleavage, or the typical α-cleavage, benzyl and phenyl cleavages, and the loss of formaldehyde and acetylene neutral molecules, as shown in Figure 6.3.S.

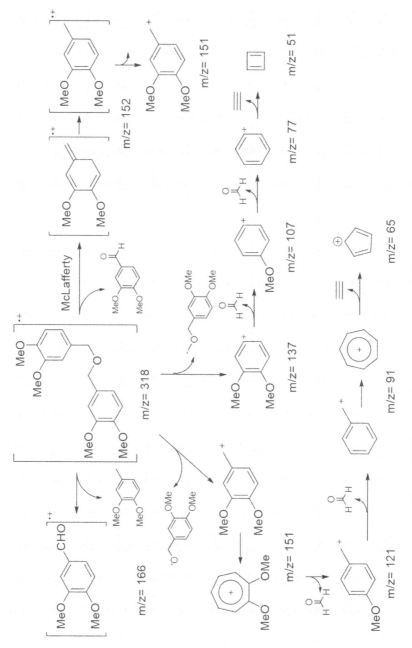

Figure 6.3.S   Main fragmentations of 3,4-dimethoxybenzyl ether.

Problem 7

The EI-MS of this compound is unusually simple. Since the molecular ion shows up at m/z 95, the compound has an odd number of nitrogen atoms. In addition, the high relative abundance of the M-1 peak (59%) indicates that there is a hydrogen that is easily lost. From the rule of thirteen, C_6H_9N might be proposed, but the 5 non-equivalent protons in the ^1H NMR indicate that the formula must be revised. The narrow doublet at 9.5 ppm in ^1H NMR and the absorptions at 1650 ($\nu_{C=O}$), 2833, and 2983 (C-H comb.) cm^{-1} in the infrared spectrum suggest that there is an aldehyde and, therefore, an oxygen in the molecule. So, after introducing an oxygen in the previous formula we find out that the molecular formula of the unknown compound 7 is C_5H_5NO, and therefore it has 4 unsaturations. The compound is likely an aldehyde and, thus it has to be a five-membered nitrogen hetero-aromatic compound in order to fit the molecular formula and unsaturations.

In the infrared spectrum, beside the aforementioned peaks, a broad band appears at 3275 cm^{-1} due to the stretching of a hydrogen bound N-H bond. The carbonyl stretching vibration appears at 1650 cm^{-1}, which is quite a low wavenumber value in concordance with a hydrogen-bonded, conjugated aldehyde.

^{13}C NMR spectra confirm the above deductions. The 5 non-equisvalent carbons are pretty deshielded and appear between 110 and 180 ppm. The relatively upshifted carbonyl carbon at 179.7 ppm fits with an aromatic aldehyde. Logically, only one carbon is missed in DEPT-135, the one bound to the formyl group that resonates at 132.8 ppm. The presence of a formyl (–CHO) group is confirmed by the abundant M-1 and M-29 peaks in the MS spectrum.

The ^1H NMR shows five one proton-signals, three complex multiplets between 6.34 and 7.17 ppm corresponding to aromatic CH protons, a doublet at 9.52 ppm (J = 0.7 Hz), and a broad singlet at 10.25 ppm. The study of the multiplets should allow us to locate the formyl group in the ring. The coupling constants in the aromatic rings are very characteristic and can be very useful when it comes to recognizing or confirming the identity of the aromatic compound since, although they are dependent on the type of ring, they do not vary substantially depending on the substitution.

In this problem, the multiplet complexity is a bit higher than expected because, in addition to the coupling of NH with all protons of the ring, the coupling with the aldehyde proton enters the scene. However, it is still possible to say with certainty in which position the pyrrole is substituted thanks to the analysis of the signals of the two least deshielded protons and the bibliographic values of pyrrole coupling constants (Pretsch et al. 2010, Abraham and Berstein 1959) (Fig. 7.1.S). The constant of 3.8 Hz present in both signals must be due to the coupling between the H3 and H4, which is the largest one in pyrrole ring, and therefore formyl group is at the C-2 position. In order to distinguish between these two protons, chemical shifts and coupling constants can be used. Owing to the presence of the formyl group, proton 3 is expected to be more deshielded. Moreover, it should show a smaller coupling constant since $^3J_{45}$ is greater than $^4J_{35}$, which is typically lower than 2 Hz in most pyrrole derivatives. On the other hand, the coupling constants between NH and H2, H3 and H4 usually are quite similar, ranging between 2.3-2.6 Hz. With this data in hand, we can say that the double doublet of doublets at 7 ppm corresponds to H3 and the doublet of triplets at 6.36 ppm is the resonance of H4. The similar strength of the NH-H4 and H4-H5 coupling is the reason why H4 signal is a doublet of triplets. The multiplet at 7.16-7.18 ppm, then, has to be assigned to H5, the proton downshifted by the neighboring nitrogen. The three similar couplings to H1, H3, H4, and the additional coupling to aldehyde proton result in the formation of an undefined multiplet, and make it impossible to calculate the coupling constants (Abraham and Berstein 1961).

Finally, the broad signal at 10.26 ppm is the resonance of the NH proton. This high chemical shift makes clear that the amine in the molecule is "special", and strongly downshifted due to the deshielding by the induced magnetic field of the heteroaromatic ring. The broadening of the signal is the result of several effects, including the observed coupling to the other aromatic protons, the intermolecular exchange and partial coupling to the quadrupolar ^{14}N. The complete assignation of the NMR signals is shown in Figure 7.1.S.

The mass spectrum confirms the proposed structure. As aromatic aldehydes usually do, the compound loses one hydrogen easily by a α-cleavage (M-1), followed by the release of carbon monoxide from the corresponding acylium ion (Fig. 7.2.S).

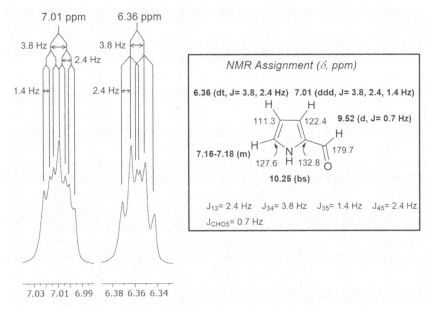

Figure 7.1.S Explanation of the 2-formylpyrrole spin system.

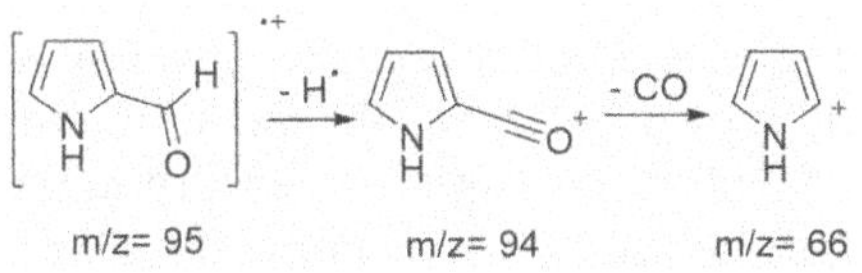

Figure 7.2.S Main fragmentations of 2-formylpyrrole.

Problem 8

The even m/z value (248) from the molecular ion of the EI-MS spectrum suggests no nitrogen or an even number of nitrogen atoms. Application of the rule of thirteen provides $C_{19}H_{20}$ as the first option, but the NMR spectra are quite simple, showing only 7 types of carbon atoms and 4 types of protons (integration $2:1:1:1$). In other words, our molecule is highly symmetric or the "heteroatomic" part is quite important. Among the few peaks that appear in the mass spectrum we observe one whose m/z is 121. This fragment would correspond to a loss of 127, which is the atomic mass of iodine. The presence of this halogen could explain the aforementioned simplicity of the NMR spectra. Other recorded ions (m/z 218 and 190) appear at M-30 and M-58 (M-30-28), thus suggesting the sequential loss of formaldehyde and carbon monoxide.

The infrared spectrum is not very helpful in this case, though the presence of C_{sp^3}-H and C_{sp^2}-H single-bond vibrations (2890-3100 cm^{-1}) along with C_{ar}-C_{ar} stretching bands at 1597 and 1498 cm^{-1} support the idea of an aromatic compound bearing aliphatic moieties. Anyway, examining the NMR spectra might be a better idea. The ^{13}C NMR spectrum shows 7 signals, six of which appear between 100 and 150 ppm, so the presence of sp^2 hydrided carbons either belonging to an aromatic ring or to various alkene fragments can be conjectured. From the DEPT-135 experiment it is clear that the two most deshielded carbons at 147.9 and 148.7 ppm are quaternary, so they could be C_{sp^2}-O or, more unlikely, C_{sp^2}-N. Although iodine is a halogen and its electronegativity is higher than that of carbon (2.66), a C_{sp^2}-I is not expected to resonate at such high chemical shift. On the contrary, the large electronic cloud of this halogen would partially compensate the deshielding due to the double bonds or the aromatic ring, thus causing an upfield shift. In fact, C_{arom}-I usually resonate between 80-100 ppm. On the other hand, it should be noted that the signal at 101.5 ppm corresponds to a methylene that could belong to a disubstituted alkene or, if not, to a very strongly deshielded alkyl methylene. Last, another quaternary carbon signal appears at 82.3 ppm. Could it be the above C_{sp^2}-I? We could think that it is an alkyne carbon, but either the alkyne is symmetric or two signals should show up in this range. Besides, the loss of 127 mass units in mass spectrum is quite indicative. In summary, there are three quaternary carbons in the molecule, one likely bound to an iodine atom, the other bound to deshielding heteroatoms, three aromatic or alkene CH, and a highly downshifted methylene. Considering what we have assumed so far, if the rule of 13 is followed again, $C_7H_5IO_2$ and

$C_7H_5IN_2$ will be obtained as possible formulas. In view of the above data, we should be more inclined to think that $C_7H_5IO_2$ is the right one. The double bond equivalents are 5, enough for 7 carbon-molecule. The molecule is likely a trisubstituted benzene, one of whose substituents is the iodine atom. Evidently, a methylenedioxy bridge attached to the benzene ring would explain both the fifth unsaturation and the high chemical shift of the methylene and quaternary aromatic carbons at 150 ppm.

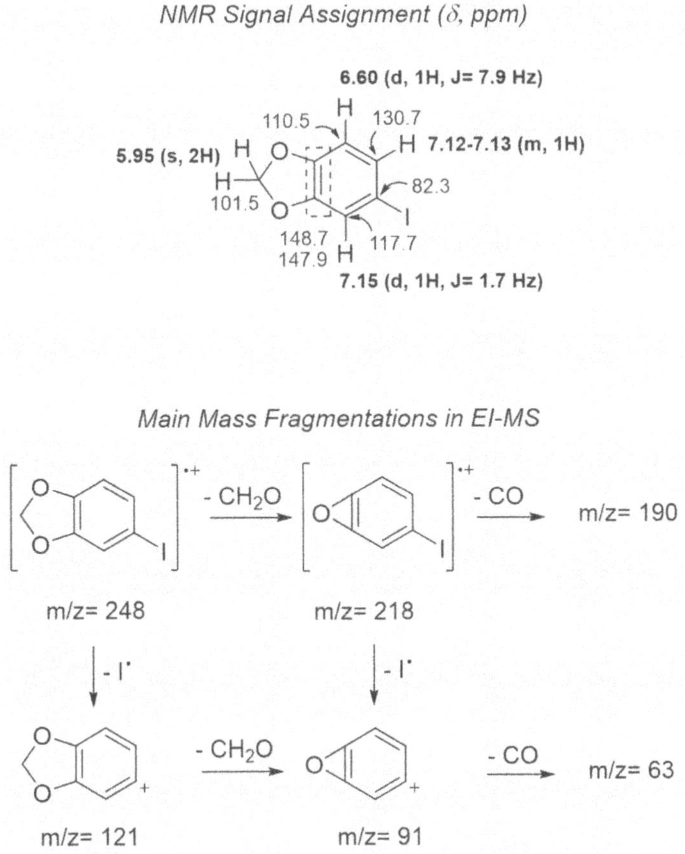

NMR Signal Assignment (δ, ppm)

6.60 (d, 1H, J= 7.9 Hz)
110.5
130.7
H 7.12-7.13 (m, 1H)
5.95 (s, 2H)
82.3
101.5
148.7
147.9
117.7
7.15 (d, 1H, J= 1.7 Hz)

Main Mass Fragmentations in EI-MS

- CH₂O → m/z= 248 → m/z= 218 - CO → m/z= 190

- I˙

m/z= 121 - CH₂O → m/z= 91 - CO → m/z= 63

Figure 8.1.S NMR signal assignment and main mass fragmentations of 5-iodo-benzo[d][1,3]dioxole.

The ^1H NMR spectrum should allow the validation of this proposal, and in case it does, also the location of the iodo substituent in the ring. A two-proton singlet at 5.95 ppm, a one-proton doublet at 6.59 ppm (J = 7.9 Hz), a one-proton multiplet between 7.12 and 7.13 ppm, and a one-proton doublet at 7.15 ppm (J = 1.7 Hz) are the signals observed in the

^{1}H NMR spectrum. The expected doublet of doublets (*ortho/meta*) is not well defined because of a 2nd order aromatic spin system. However, the well resolved *ortho* and *meta* doublets point to 5-iodobenzo[*d*][1,3]dioxole (Fig. 8.1.S). In spite of being protons bound to a C$_{sp^3}$ carbon, the singlet of the methylene bridge protons is very downfield (6.6 ppm) because of the deshielding effected by the two oxygen atoms and the aromatic ring.

As shown in Figure 8.1.S, this functionality also explains the sequential loss of 30 mass units (formaldehyde) and 28 (carbon monoxide) observed in the mass spectrum (Stévigny et al. 2004).

Problem 9

From mass spectrum, the m/z of the molecular ion (158) is deduced, and therefore there is no nitrogen or an even number of nitrogen atoms. There is likely a methoxy group (OMe) in the molecule since the loss of 31 mass units (M-31) is the base peak of the spectrum. The loss of methyl (M-15) appears too, with a relative abundance of 12%. Moreover, an abundant peak at m/z 99 (M-59) is indicative of the consecutive loss of OMe and CO (M-31-28). The suspicion of the presence of a methoxycarbonyl unit (–COOMe) is reinforced by observing the peak at m/z 59 (MeOOC$^+$).

The intense absorption at 1741 cm^{-1} in the infrared spectrum evidences the presence of a carbonyl group. The relatively low wavenumber value points to an aldehyde carbonyl or, maybe, a α,β-unsaturated ester carbonyl. In this regard, the symmetric and asymmetric stretching of the C-H formyl group are not observed, but there is an absorption at 1639 cm^{-1} which may be due to the stretch of a C=C double bond. In addition, the weak band just above 3000 cm^{-1} is also diagnostic of unsaturation (alkenyl C-H stretch).

The ^{13}C NMR spectrum shows the resonances of seven non-equivalents carbon. A methylene carbon (see the phase of the peak in DEPT-135 experiment) that resonates at 37.3 ppm suggests the presence of a carbonyl or a double bound in its chemical environment that effects a light deshielding. Two peaks appearing a little more downfield (51 ppm) might correspond to two methyl groups bound to a heteroatom, which is an option in concordance with the loss of methoxy groups observed in the mass spectrum. Besides these signals, we can see a strongly deshielded methylene that resonates at 128.9 ppm, which is, obviously, an alkene carbon, and three quaternary carbons-one at 133.6 ppm, which is probably the second carbon of the disubstituted alkene, and two signals further downfield, exactly at 166.4 and 170 ppm, which suggest the presence of two ester carbonyls. The information provided by the ^{13}C NMR spectrum and the m/z of the molecular ion allow us to propose the molecular formula of the molecule using the rule of thirteen. The molecular formula of unknown compound 9 is therefore $C_7H_{10}O_4$, and has three double bond equivalents on account of the two carbonyls and the alkene.

Thanks to the ^{1}H NMR, the pieces of the puzzle can be put together. The two-proton singlet at 3.34 ppm must be the resonance of the aliphatic methylene that appears at 37.3 ppm in the ^{13}C NMR spectrum. The chemical shift is quite high and indicates that, either this methylene is next to a heteroatom and a quaternary carbon, or it is deshielded by the anisotropy of the two carbonyls, or by one of the carbonyl groups and the alkene. The two three-proton singlets at 3.7 ppm confirm the presence of the two methoxycarbonyl groups. Finally, the geminal alkene protons resonate at 5.71 and 6.32 ppm, and show a coupling constant of 1.1 Hz. Dimethyl 2-methylenesuccinate (dimethyl itaconate) fits all the above experimental data. Both alkene protons are expected to be deshielded by the field generated by the double bond and the electron withdrawing –COOMe group. However, one of the protons is considerably more downfield shifted

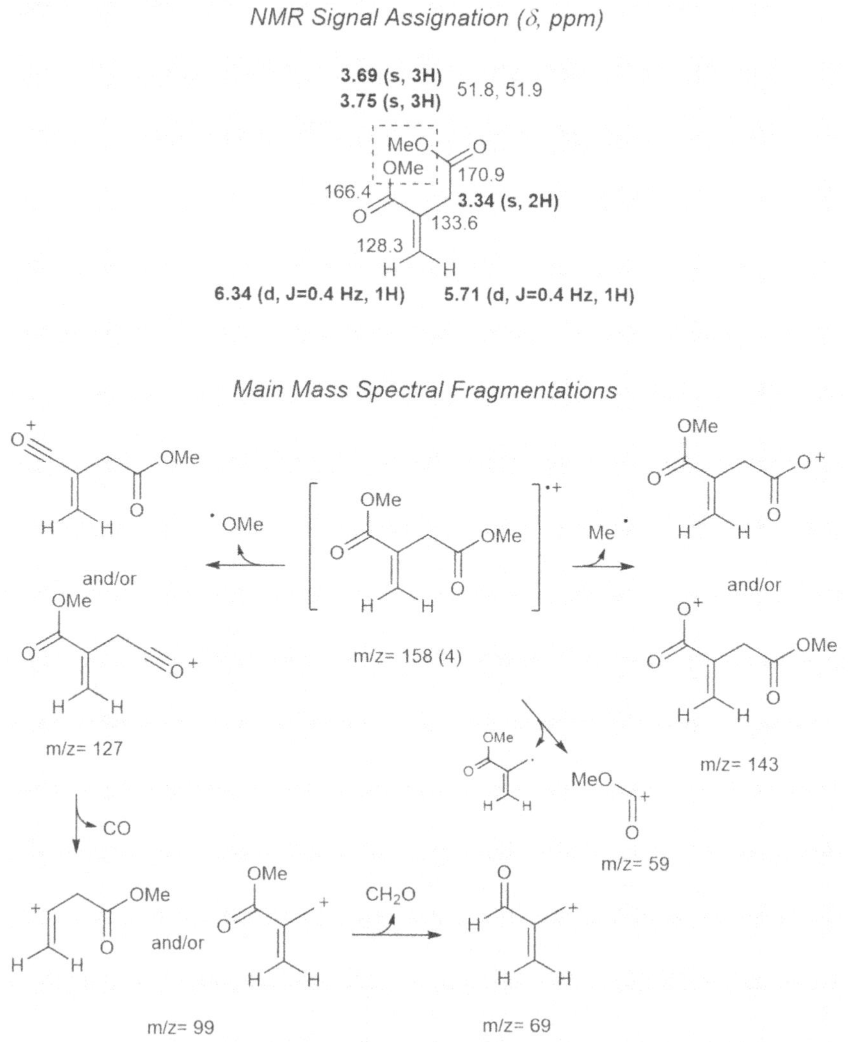

NMR Signal Assignation (δ, ppm)

Main Mass Spectral Fragmentations

Figure 9.1.S   NMR signal assignment and main mass fragmentations observed in the EI-MS of dimethyl itaconate.

than the other. This fact is crucial to assign both signals. The carbonyl of the ester has two effects. It not only withdraws electron density from the β-carbon and deshields the proton bound to it, but also generates, due to the magnetic anisotropy, a strongly deshielding region in the plane of the carbonyl group. Both effects explain the higher chemical shift observed for the "*cis*" alkene proton (Fig. 9.1.S).

Finally, as mentioned before, α-cleavages and subsequent decarbonylations explain most of the peaks observed in the mass spectrum.

Problem 10

It is not unusual to mix the molecular ion peak (M^+) and that from the loss of hydrogen atom ($M-1$). Such mistakes can be avoided considering the expected abundance of the $M+1$ peak, normally due to ^{13}C isotope. In this case, from the mass spectrum we know that the m/z of the molecular ion is 98, and that this ion loses hydrogen and methyl radical easily since the peaks of the corresponding fragments ($M-1 \rightarrow m/z\ 97$ and $M-15 \rightarrow m/z\ 83$) are the ones with highest relative abundance.

The infrared spectrum of this compound exhibits diagnostic bands at 1440 and 1538 cm^{-1} ($\nu_{Csp^2-Csp^2}$), the overtones between 1600 and 1800 cm^{-1}, and the C-H bond stretching bands at 3049, 3072, and 3106 cm^{-1}, all suggesting the presence of an aromatic ring. In addition, the ν_{Csp^3-H} bands at 2862, 2922, and 2950 cm^{-1} are indicative of the presence of an alkyl moiety.

The peaks between 120 and 140 ppm in the ^{13}C NMR spectrum confirm the presence of an aromatic ring. This is likely a five-membered heteroaromatic ring since there are three non-equivalent CH (123.1, 125.2, and 126.9 ppm) and only a quaternary carbon (139.6 ppm). The chemical shift of the most shielded signal, 15.1 ppm, induces us to think that the ring substituent is a methyl group. Besides, mass spectrum shows that the loss of methyl happens without difficulty after the generation of the molecular ion. With all this information and the molecular mass we can effortlessly guess that the molecule is a methylthiophene.

The ^1H NMR shows three-proton multiplets between 6.7 and 7.1 ppm, which evidently correspond to the resonances of the aromatic protons, and a three-proton doublet. The fact that the signal of the methyl resonance splits into two lines indicates that these protons are coupled with a non-equivalent proton, which can only be a proton of the ring. Benzylic couplings are not unusual in aromatic structures, but the coupling constants are often too small for the doublet to be resolved, and usually result only in the widening of the signals. In this case, a long-range coupling constant of 1 Hz is clearly observed. In addition, the spin-spin coupling patterns and the coupling constants of the aromatic signals allow us to confirm the identity of the heteroaromatic ring, locate the substituent, and assign all the proton resonances unambiguously. The spectrum shows a doublet of doublets at 7.08 ppm (J = 5.1, 1.1 Hz), a doublet of doublets at 6.90 ppm (J = 5.1 and 3.4 Hz), and what looks like an apparent doublet of triplets at 6.76 ppm (J = 3.3 Hz and 1.1 Hz). The presence of the 5-Hz coupling constant is a hint that the compound is a thiophene since the constants are smaller in the case of furans or pyrroles, being the highest in these rings around 3.5 Hz. This coupling constant is the ^3J between the protons at positions 2 and 3 in thiophene. Taking into account the common values of the rest of the coupling constants in this ring ($^3J_{H3-H4}$ ~3.5 Hz, $^4J_{H2-H3}$ ~1.0 Hz, $^4J_{H3-H4}$ ~2.8 Hz) and the observed signals, we can conclude that the compound is 2-methylthiophene. The ^4J benzylic coupling is responsible for the coupling pattern observed in the signal of H3, which *a priori* is expected to be a doublet of doublets (see Fig. 10.1.S).

Along with the abovementioned benzylic and α-type cleavages, loss of acetylene and more specific fragmentations of thiophenes explain the MS spectrum (Fig. 10.1.S).

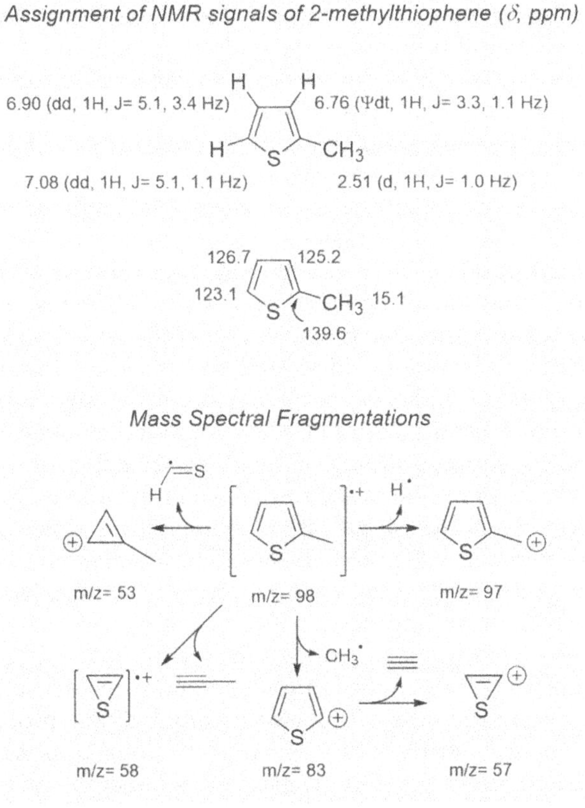

Assignment of NMR signals of 2-methylthiophene (δ, ppm)

Mass Spectral Fragmentations

Figure 10.1.S NMR signal assignment and main mass fragmentations of the 2-methylthiophene.

Problem 11

We have the accurate mass from the high resolution mass spectrum in this case. There are calculators that provide the most probable molecular formulas from the monoisotopic mass (www.chemcalc.org, Patiny and Borel 2013). From given accurate molecular mass, the formula whose mass is closest to the experimental value is C_4H_8O. Therefore, the compound is a little oxygen molecule with a hydrogen deficiency index of 1. In the LRMS, the fragment formed by loss of 17 mass units (m/z = 55, 10%) and the ion whose m/z value is 45 (30%) suggest that the molecule is an aliphatic alcohol. The relative abundance of the molecular ion in the LRMS is quite low, which is something usual in alcohols since loss of hydroxyl group or dehydration take place quite easily. In addition, M-15 (M-CH_3) is the base peak (m/z 57), thus suggesting that both hydroxy and methyl groups are bound to the same carbon (α-cleavage).

The ^{13}C NMR spectrum shows two carbons strongly deshielded, one methylene (DEPT-135) that resonates at 113. 4 ppm, and a CH signal further downfield (142.1 ppm), making clear that the unsaturation of the molecule derives from a monosubstituted C=C double bond. The resonances of the hydrocarbon substituent appear at 22.9 and 68.7 ppm. The CH bound to the heteroatom will be downfield shifted, so it is clearly the signal at 68.7 ppm. Since neither of them is a methylene, we can easily deduce that the compound is not a terminal alcohol. The compound must be 3-buten-2-ol.

The ^1H NMR shows 6 signals. Upfield we see a three-proton doublet (J = 6.4 Hz) at 1.26 ppm due to the resonance of the methyl group, and at the OH peak, a one-proton singlet at 1.87 ppm. Further downfield, there is a multiplet centered at 4.28 ppm that should be assigned to the CH bound to the hydroxyl group. The signal is an apparent quintuplet of apparent triplets, which is a coupling pattern as a result of two vicinal couplings (one with the methyl protons and the other with the neighboring alkene proton, ~6 Hz) and two allylic couplings similar to each other (~1 Hz) (Fig. 11.1.S). The resonances of the two geminal alkene protons are two doublets of apparent triplets at 5.04 and 5.19 ppm, thus meaning that not only the two allylic coupling constants, but also the alkene geminal constants, are of similar magnitude. Thanks to the *trans* and *cis* coupling constants (17.4 Hz and 10.4 Hz, respectively), the alkene signals can be unequivocally assigned. The most deshielded proton is the alkene CH, whose resonance is a doublet of doublet of doublets because of the three different vicinal couplings (17.2, 10.4, and 5.9 Hz) (Fig. 11.1.S).

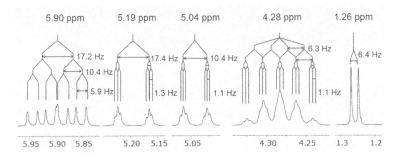

Figure 11.1.S Analysis of ^{1}H NMR coupling constants of unknown compound **11**.

Finally, the fragmentation pattern on the EI-MS spectrum is based on α-cleavages, which lead to the loss of hydrogen atom and methyl, vinyl and hydroxyl radicals, and subsequent loss of ethylene (Fig. 11.2.S).

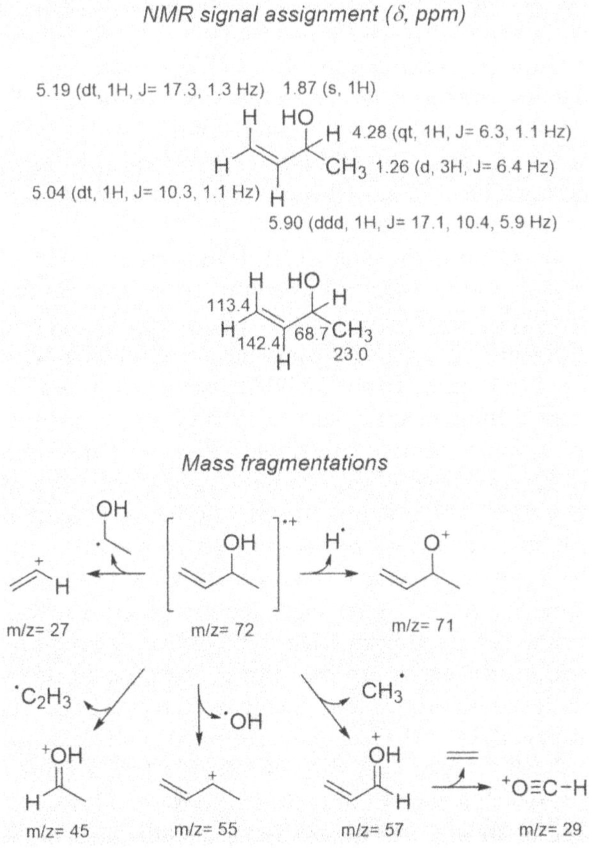

NMR signal assignment (δ, ppm)

5.19 (dt, 1H, J= 17.3, 1.3 Hz)   1.87 (s, 1H)

H 4.28 (qt, 1H, J= 6.3, 1.1 Hz)
CH$_3$ 1.26 (d, 3H, J= 6.4 Hz)

5.04 (dt, 1H, J= 10.3, 1.1 Hz)

5.90 (ddd, 1H, J= 17.1, 10.4, 5.9 Hz)

Mass fragmentations

m/z= 27   m/z= 72   m/z= 71

m/z= 45   m/z= 55   m/z= 57   m/z= 29

Figure 11.2.S NMR signal assignment and mass fragmentations of 3-buten-2-ol.

Problem 12

The odd m/z value of the molecular ion (m/z = 121) points out to a nitrogen compound. The relative abundance of the M$^+$ peak accounts for the stability of the molecular ion. The loss of methyl generates the base peak of the spectrum (m/z = 106) and the subsequent loss of CO (loss of 28 mass units, M-15-28, m/z 78) gives rise to the second most abundant fragment, thus suggesting the presence of an acetyl group in the molecule.

The infrared spectrum shows an intense absorption at 1688 cm^{-1}, which is probably the stretching vibration of a ketone carbonyl. The frequency is a bit low, indicating a conjugated ketone. Besides, the absorption at 1585 cm^{-1} could be the stretching of the C-C of an aromatic ring. Although the wide band at around 3370 cm^{-1} looks like an O-H vibration, it probably corresponds to the presence of moisture in the sample, because no loss of 17 and/or 18 mass units is observed in the MS. Another possible source of the above wide band would be an overtone of the C=O stretch, commonly found in methyl ketones. In addition, weak bands at 3086, 3050, and 3006 cm^{-1} (potential ν_{Csp^2-H} stretching) and at 2971 cm^{-1} (potential ν_{Csp^2-H} vibration) are also observed.

Taking into account the mass of the molecular ion, the fact that it is a nitrogen compound, and the presence of at least an oxygen in the molecule, C_7H_7NO can be proposed as the molecular formula by using the rule of thirteen. The seven signals of non-equivalent carbons in ^{13}C NMR and the integration from ^1H NMR (seven or a multiple of seven are the protons of the molecule) support the proposed formula. The high unsaturation index (5) for a relatively small organic molecule is usually explained by the presence of an aromatic ring, and just a glance at ^{13}C NMR is enough to confirm this. There is only one highfield signal at 26.5 ppm, which is probably the resonance of a methyl group, lightly deshielded by a carbonyl group or an aromatic ring. The rest of the signals are further downfield, in the aromatic region, except for the quaternary carbon resonating at 196.5 ppm, which obviously belongs to the ketone carbonyl already deduced from MS and IR spectra. With regard to the signals of four tertiary carbons and one quaternary carbon (DEPT-135) that resonate between 123.4 and 153.2 ppm, they are in all likelihood aromatic carbons, but the high chemical shift of two CH that appear at 149.6 and 153.2 ppm is noticeable. It is common to find downfield shifted quaternary aromatic carbons attached to heteroatoms, usually oxygen, but they are tertiary carbons in this case. This fact provides an important clue about the nature of the aromatic ring. The presence of a heteroatom in the ring would explain the strong deshielding of these two carbons. Considering that the compound contains nitrogen and that there is a single quaternary carbon, one reaches the conclusion that the compound is a monoacylated pyridine.

Inspection of the ^{1}H NMR allows us to deduce where the acetyl group is located. There are 5 signals in the spectrum - a three proton singlet at 2.63 ppm, expected chemical shift for the methyl of an acetyl pyridine, and the 4 one-proton signals of the aromatic protons: a doublet of double doublets at 7.42 ppm (J = 8.0, 4.8, 0.8 Hz), a doublet of apparent triplets at 8.22 ppm (J = 8.0, 2.0 Hz), a doublet of doublets at 8.77 ppm (J = 4.8, 1.7 Hz), and a doublet at 9.51 ppm (J = 2.1 Hz). Coupling constants in pyridine derivatives are quite characteristic. The most strongly coupled protons are H3 and H4 (or H4 and H5), with the coupling constant between 7 and 9 Hz. The other vicinal constant (between protons H2 and H3, or H5 and H6) is a bit smaller, ranging between 4 and 6 Hz, and finally the coupling constants through four bonds have magnitudes between 0 and 2.5 Hz. As no multiplet with two 7-9 Hz couplings is observed, it is easy to deduce that the acetyl group is on carbon 3, and therefore the compound is 3-acetylpyridine. The magnitude of the coupling constants facilitates the assignation of the heteroaromatic proton signals. Besides, the most deshielded resonances are the ones next to the nitrogen. The doublet with a little coupling constant is the resonance of H2, which is next to both nitrogen and acetyl group and is not strongly coupled. The dd at 8.77 ppm is the resonance of H6, coupled to a vicinal proton (H5), as well as to H4. The doublet of apparent triplets at 8.22 ppm, which shows only a vicinal coupling and is considerably deshielded, will be the signal of the proton *ortho* with respect to the acetyl group and *para* with respect to the nitrogen atom. Finally, the most shielded proton will be H5, whose signal is a ddd with two vicinal coupling constants. Regarding the carbon signals, in the absence of other data, the complete assignment must be done based on the theoretical chemical shifts calculated using any of the available tools (Fig. 12.1.S).

In addition to the abovementioned α-cleavage and decarbonylation leading to the pyridinylium ion, the loss of HCN by this cation, which is a typical fragmentation of this heteroaromatic ring, can be observed in this EI-MS spectrum (Fig. 12.1.S).

Assigment of the NMR signals (δ, ppm)   Main mass fragmentations

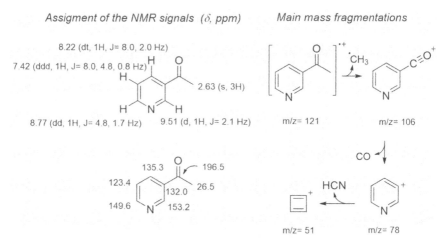

Figure 12.1.S NMR signal assignment and EI-MS fragmentations of 3-acetylpyridine.

Problem 13

Two peaks separated by 2 mass units with a peak height ratio of $1:1$ (m/z 194 and m/z 196) indicate that the molecule contains a bromine atom. Monoisotopic mass is even, so there is no nitrogen in the molecule, or at least not an odd number of nitrogen atoms. The loss of 45 and 73 mass units suggests the presence of an ethoxycarbonyl group (M-OEt and M-COOEt). This idea is supported by the infrared spectrum, which shows an intense absorption at 1731 cm^{-1}.

From the rule of thirteen and with the presence of two oxygens and a bromine in mind, $C_6H_{11}BrO_2$ can be proposed as the molecular formula of this unknown compound whose unsaturation index would be one.

The ^{13}C NMR shows six signals, making clear that none of the carbons are equivalent. Except for the strongly deshielded signal at 172.6 ppm belonging to the ester carbonyl, all the resonances appear before the triplet of deuterated chloroform, as expected from a monounsaturated compound. One of the carbons resonates at a high frequency to be an aliphatic carbon (60.7 ppm), which indicates that it is attached to an electron-withdrawing group or atom, the ester group, obviously. The rest of the carbons are further highfield. One is specially shielded and its resonance appears at 14.3 ppm, suggesting that is a terminal methyl. The other three appear a bit downfield, at 27.9, 32.6, and 32.9 ppm, surely due to the deshielding effect of the ester carbonyl and the bromine atom. In this case there is no DEPT experiment to determine multiplicity of carbon atom substitution with hydrogens, but we do have a HSQC two-dimensional spectrum which, after studying the ^1H NMR spectrum, should provide this information as well as the assignment of the carbon signals.

The ^1H NMR shows 5 types of protons. There is a three-proton signal that shows up at 1.25 ppm, a triplet with J = 7.1 Hz, and the rest are two-proton multiplets; a quintuplet at 2.16 ppm (J = 6.8 Hz), a triplet at 2.48 ppm (J = 7.2 ppm), another triplet at 3.46 ppm (J = 6.5 Hz) and, finally, a quadruplet at 4.13 ppm (J = 7.1 Hz). All the coupling constants are similar and around 7 Hz, which is a typical vicinal coupling constant value for an open-chain compound with free rotation. This fact explains the appearance of a quintuplet, although there are not four equivalent protons in the molecule. The signal of an internal methylene (if the two protons of each methylene are equivalents) should be a triplet of triplets, but the overlapping of the lines due to the similar coupling constants results in an apparent quintuplet. The coupling pattern indicates that there is an ethyl

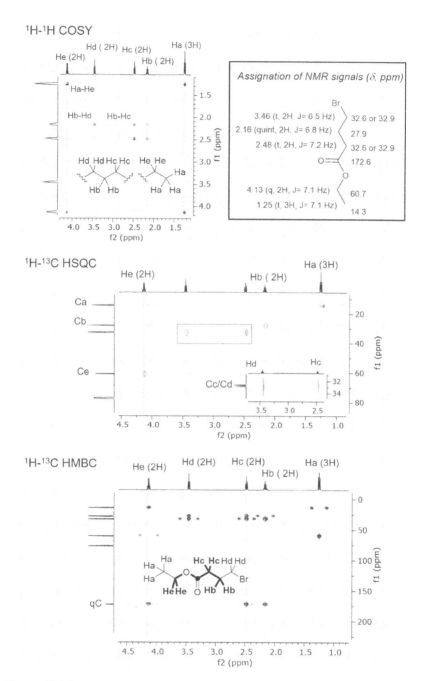

Figure 13.1.S Homonuclear and heteronuclear correlations found for ethyl 4-bromobutyrate. NMR signal assignment.

group and a three-methylene chain. The methylene of the ethyl group, whose quadruplet appears at 4.13 ppm, is quite deshielded, so it must be the one bound to the oxygen of the ester group. Evidently, the bromine atom has to be located at the other end of the chain in other to be a three-methylene chain, so the unknown compound is ethyl 4-bromobutyrate. After the examination of the chemical shifts and the analysis of the multiplets, the assignment of the ^{1}H NMR signals is straightforward, since the second most deshielded protons must be those next to the bromine atom, and the coupling pattern indicates the connection of the chain methylenes. Anyway, COSY, HSQC, and HMBC 2D-NMR experiments are available in this case, so an unequivocal assignment of the signals can be done. As shown in Figure 13.1.S, the COSY experiment confirms the proposed spin systems, and thanks to HSQC and HMBC experiments, the carbonyl group is correlated with its neighboring methylenes.

Finally, we observe that the proposed structure is consistent with all the spectral data in EI-MS. The loss of bromoethylene via McLafferty rearrangement yields the fragment with m/z 88 as the base peak of the spectrum. This fragmentation is accompanied by α-bond cleavages with subsequent loss of CO, other McLafferty rearrangements, and loss of bromine atom (Fig. 13.2.S).

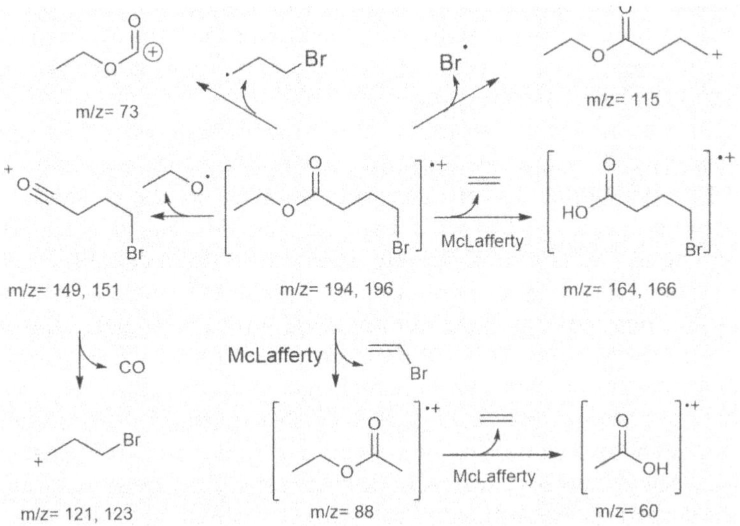

Figure 13.2.S Main mass fragmentations of ethyl 4-bromobutyrate.

Problem 14

The peak at the highest m/z value of the mass spectrum does not seem to correspond to the molecular ion since no logical loss explains the base peak (m/z 59) of the spectrum. The difference in mass between the latter and the next most abundant peak (m/z 41) is 18, which might be attributed to the loss of water, thus suggesting the presence of an OH group in the molecule. The appearance of an abundant peak at m/z 31 (CH_3O^+), a fragment of which is usually present in the mass spectrum of alcohols, supports this hypothesis.

In this regard, absorptions are visible in the infrared spectrum further beyond 3000 cm^{-1}. Thus, a broad band that extends from about 3200 cm^{-1} to 3500 cm^{-1} overlaps with a narrower band at 3300 cm^{-1}, suggesting the presence of O-H or N-H bonds, or even a terminal alkyne whose C-H stretching tension shows up in this region. In the C≡C bond stretch region, only a tiny absorption around 2200 cm^{-1} can be found, although this fact does not rule out the presence of an alkynyl moiety, since their infrared absorptions are usually very weak. In fact, if the mass spectrum is carefully examined, a peak at m/z 39, which would correspond to propargylic cation, can be found.

With no molecular formula, the examination of NMR spectra becomes crucial. Six resonances appear in ^{13}C NMR and there is no signal beyond 80 ppm, so we can forget double bonds, aromatic rings, and carbonyl groups. However, three carbon are quite deshielded and resonate at 70.7, 71.3, and 81.0 ppm. These high chemical shifts suggest the presence of the abovementioned sp-hybridized carbons and/or the presence of heteroatoms that strongly deshield these nuclei. All of these signals are positive peaks in DEPT experiment, thus probably implying that they are directly bound to one hydrogen and, therefore, the presence of a carbon-carbon triple bond could be discarded. However, we must note the much lower intensity of the carbons at 81.0 and 70.7 ppm, which could be related with an unusual constant between these carbons and coupled protons (see Solution to Problem 4 for more details). Besides these signals, we find a resonance at 9.9 ppm, the signal of a terminal methyl carbon, and two methylene carbons at 26.9 and 29.1 ppm. The cumulative mass of these carbon fragments (one CH_3, two CH_2, and three CH) is 82 mass units. If there were an oxygen in the molecule (something that seems quite probable in view of the chemical shifts of some of the carbons), the cumulative mass would add up to of 98, which is a value 39 mass units heavier than the base peak of the spectrum. This reasoning points back to the hypothesis of the presence of a terminal alkyne.

The ^1H NMR confirms the presence of 10 protons in the molecule. A three-proton triplet at 0.93 ppm with a typical vicinal J value (7.5 ppm) and the two-proton complex multiplet between 1.37 and 1.70 ppm probably belong to a relatively deshielded ethyl group. A bit downfield (2.02 ppm), we find a one-proton triplet that shows a meaningful coupling constant (2.7 Hz). This coupling is quite little for a carbon chain with free rotation to be a vicinal coupling, thus indicating a long range coupling. The chemical shift and the coupling constant should remind us of an acetylenic hydrogen whose signal is split by the methylene hydrogens across the triple bond. Next to this signal, a one-proton broad singlet at 2.18 ppm looks like the resonance of an OH proton. The coupling pattern of two one-proton doublets of double doublets, one at 2.30 ppm (J = 16.6, 6.7, 2.7 ppm) and the other at 2.40 ppm (J = 16.8, 4.9, 2.7 ppm), indicates that they are two geminally coupled diastereotopic protons (J = 16 Hz) probably coupled to the acetylenic proton (J = 2.7 Hz). The chemical shifts indicate that these protons are slightly deshielded. As ^{13}C NMR show no carbonyl group or Csp^2 in the molecule, the deshielding should be due to the presence of an electron-withdrawing heteroatom or the neighboring sp carbon. The effect of the suggested oxygen is clearly seen in the deshielding of the proton around 3.65 ppm. The signal is a multiplet that does not allow further analysis.

Close examination of the HSQC (Fig. 14.1.S) confirms the diastereotopicity of the protons at 2.3 and 2.4 ppm, which are both correlated to the carbon at 26.9 ppm. On the other hand, both carbons at 70.7 ppm and 81.1 ppm are coupled with the acetylenic proton. As we commented in Problem 4, $^1J_{CH}$ and $^2J_{CH}$ values are often unusually high in alkynes - a fact that frequently results in the appearance of unexpected signals in DEPT experiments or unexpected correlations in HSQC experiments. Finally, there is no correlation between the proton at 2 ppm and any of the carbons, so we can conclude that this hydrogen is bound to heteroatom, as its ^1H NMR signal suggests.

Once the direct connectivities between proton and carbon nuclei are understood, and the long-range coupling of the alkyne proton is cleared up, COSY experiment allows the unambiguous determination of the homonuclear spin system of the molecule so that a proposal for the structure of unknown compound **14** can be made. Thus, the data so far points to 5-hexyn-3-ol, whose NMR signal assignment is shown in Figure 14.1.S.

The coupling constants between diastereotopic hydrogens at carbon 4 (H_e and H_f) and the neighboring hydrogen (H_g) are a bit smaller. The reason can be the influence of the hydroxyl group, because the presence of electronegative substituents causes smaller J values, in particular in the

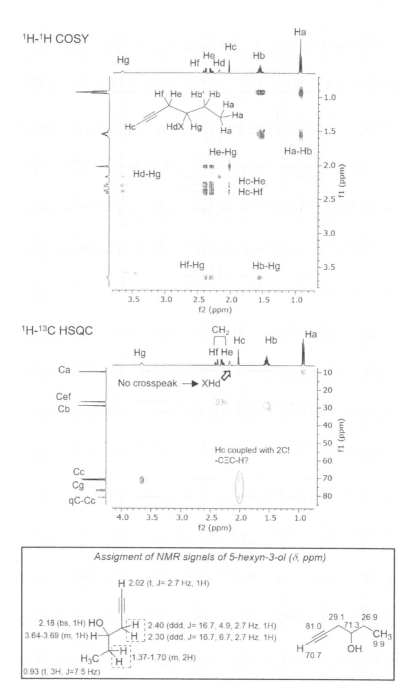

Figure 14.1.S   Direct correlations found in HSQC and COSY experiments. NMR signal assignment for 5-hexyn-3-ol.

case of hydrogens on carbon bound to oxygen substituents (Bothner-By 1965).

Interestingly, it is worth mentioning that the value of J is different for these diastereotopic protons (6.7 Hz versus 4.9 Hz). The dependence of the coupling constants on torsional or dihedral angles is known, and it has been quantified by Karplus equation, first, and other models later (Karplus 1963, Wasylishen and Schafer 1973, Haasnoot et al. 1980, Tvaroska and Taravel 1995, Balacco 1996, Aydin and Günther 1990). This equation shows that vicinal H-H couplings will be maximal for protons with 180° and 0° dihedral angles because orbital overlap is optimal in these cases. In open-chain systems with free rotation, the observed vicinal coupling constant is the average value of the coupling constants at stable conformations, often resulting in a J between 6 and 8 Hz. In our case, the calculated coupling constant would be a bit lower (estimated using the online tool developed by Haasnot et al. (Haasnoot et al. 1980)[6]. The two different J values obtained from NMR spectrum evidence a slight preference for the minimum energy antiperiplanar conformation, thus providing a little higher $^3J_{H\text{-}H}$ for the diastereotopic proton antipleriplanar to the hydrogen at C4 (Fig. 14.2.S).

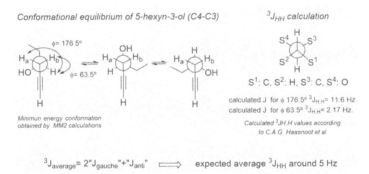

Conformational equilibrium of 5-hexyn-3-ol (C4-C3) $^3J_{HH}$ calculation

$\phi= 176.5°$
$\phi= 63.5°$

S^1: C, S^2: H, S^3: C, S^4: O

calculated J for ϕ 176.5° $^3J_{H.H}$= 11.6 Hz.
calculated J for ϕ 63.5° $^3J_{H.H}$= 2.17 Hz.

Minimun energy conformation obtained by MM2 calculations

Calculated $^3J_{H.H}$ values according to C.A.G. Haasnoot et al.

$^3J_{average}= 2"J_{gauche}"+"J_{anti}"$ ⟹ expected average $^3J_{HH}$ around 5 Hz

Figure 14.2.S Explanation of the vicinal coupling of diastereotopic hydrogens at C4 of 5-hexyn-3-ol.

Finally, despite the missing molecular ion, the mass spectrum confirms the proposed structure. The presence of a m/z 31 peak is usually found in the spectrum of primary alcohols, but secondary alcohols can undergo an "onium" rearrangement that results in the fragment at m/z 31. Other α-cleavages and propargyl fragmentations as well as dehydration processes are depicted in Figure 14.3.S.

---

[6] http://www.stenutz.eu/conf/haasnoot.php Other online tools for coupling constant calculation can be found at: https://www.spectroscopynow.com and http://www.inmr. net/sweetj.html.

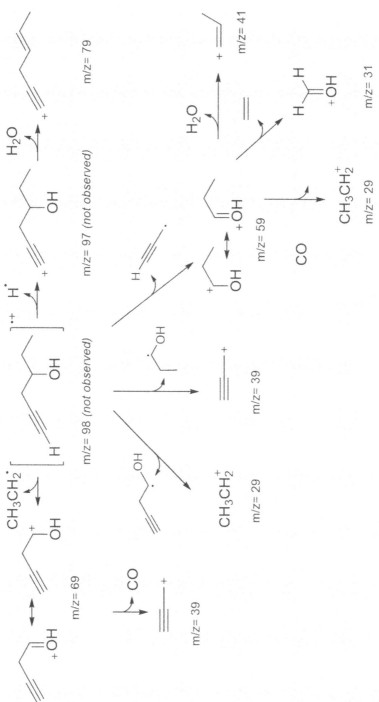

Figure 14.3.S   Mass fragmentation pattern of 5-hexyn-3-ol.

Problem 15

The pattern of the molecular ion (m/z 218/220, 41%:39%) indicates that there is a bromine atom in the molecule and no nitrogen. As a result of the sequential loss of 17 and 28 mass units, the peaks at m/z 201/203 and m/z 175/177 suggest we are dealing with a carboxylic acid that easily loses hydroxyl radical and carbon monoxide. In fact, the fragment after decarboxylation and loss of hydrogen bromine is the base peak of the spectrum (m/z 94). From this data, $C_8H_{11}BrO_2$ can be proposed as the molecular formula, so it is a compound with four unsaturations.

The broad band centered at 3393 cm^{-1} and the absorption at 1702 cm^{-1} in the infrared spectrum confirms the presence of the carboxy group. The width of the band and the absorption at 1650 cm^{-1} make us think of an aromatic compound.

On the other hand, there are thirteen peaks in the ^{13}C NMR spectrum, more peaks than initially expected and all of them between 115 and 167 ppm, in the region of sp^2 carbons. However, one can easily notice that with the exception of the signal at 166 ppm, which likely corresponds to the carboxyl carbon, all the signals are duplicated, being six doublets (Fig. 15.1.S). As we know, in a broadband proton-decoupled spectrum, the carbons are coupled with active nuclei other than proton. The measured coupling constants make clear that the nucleus is a fluorine. Therefore, we must revise the molecular formula by inserting this element. The new formula is $C_7H_4BrFO_2$, so the unsaturation index is 5. Those unsaturations are probably derived from an arene ring (4) and the carboxy group (1). The magnitude of the coupling constants from ^{13}C NMR allows an easy determination of the ring-substitution pattern and the assignment of the carbon signals, as shown in Figure 15.1.S. The doublet at 160.0 ppm with the strong coupling (J = 248.1 Hz) is the carbon directly bound to fluorine; the doublets at 118.8 and 117.4 ppm with J around 20 Hz are the carbons at *ortho* positions; the signals at 134.9 y 133.8 ppm are the resonances of the *meta* carbons (J ~7 Hz); and the doublet at 114.9 ppm is the *para* carbon (J ~3.3 Hz). This last signal (165.8 ppm) and the one at 133.8 ppm do not appear in the DEPT experiment, so the bromo and carboxy substituents are probably *meta* and *para* with respect to the fluoro group.

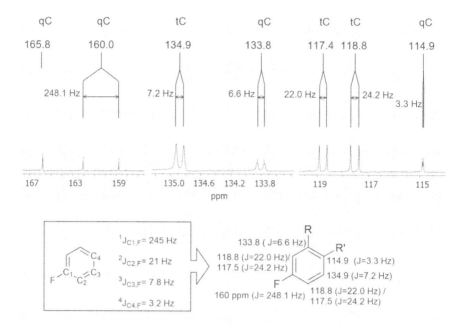

Figure 15.1.S   Measurement of ^{13}C-^{19}F coupling constants in ^{13}C NMR spectrum of unknown compound **15**.

The ^1H NMR shows only three-proton signals in the aromatic region probably because the acid proton is strongly downshifted and it resonates beyond 9.00 ppm. Due to the first-order spin system of the aromatic protons, they can be readily analyzed, showing a triplet of doublets at 7.33 ppm (J = 8.5, 3.2 Hz) and two doublets of doublets a bit downfield-one at 7.58 ppm (J = 8.8, 5.1 Hz) and the other at 7.75 ppm (J = 8.8, 5.1 Hz). Again, the J_{HF} and J_{HH} coupling constants are valuable tools to assist in the assignment of the signals, as displayed in Figure 15.2.S. The similar heteronuclear and homonuclear ^3J causes the resonance of the proton with two *ortho* couplings (H$_o$) to split into an apparent triplet of doublets. On the other hand, the *meta* coupling to the fluorine (J = 5.1 Hz) assigns the signal at 7. 75 ppm to H$_m$.

The only missing data is the location of the bromo and carboxy substituents with respect to the fluoro one. The theoretical chemical shifts suggest 2-bromo-5-fluorobenzoic acid, because the triplet of doublets signal is expected to appear further downshifted in the case of 2-bromo-4-fluorobenzoic acid. In order to solve this problem, the correlations of the carboxylic carbon in the HMBS experiment should be analyzed

(Fig. 15.3.S). As usual, direct connectivities are present as large doublets due to $^1J_{CH}$ data that can be used to unambiguously assign the signal of the two carbons next to the fluoro substituent. The two- and three-bond correlations of the carboxylic carbon should provide the needed information to locate this group in the ring. However, we can see that this carbon is correlated with two different protons at 7.58 ppm and 7.75 ppm. In both probable structures, a three bond-coupling interaction and a four-bond coupling interaction (four-bond correlations are sometimes observed in conjugated systems) could take place with these protons, so no definitive evidence can be obtained. The only visible difference we can find is the size of the observed peaks, which usually depends on the size of the coupling constant. Since $^4J_{CH}$ constants are remarkably smaller than $^3J_{CH}$, we can propose that the proton at 7.58 ppm is the one correlated to the carboxyl carbon over a three-bond interaction. This proposal is consistent with the theoretical chemical shifts of both proton and carbon nuclei. So, our molecule is 2-bromo-5-fluorobenzoic acid.

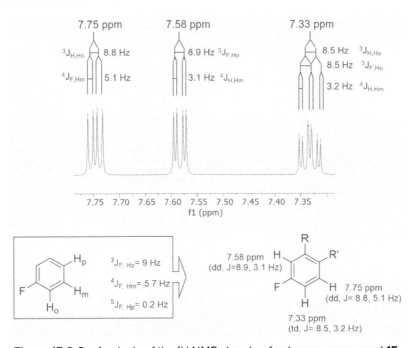

Figure 15.2.S   Analysis of the ^{1}H NMR signals of unknown compound **15**.

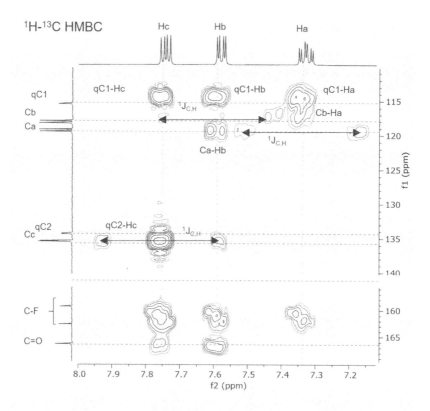

Possible location of the carboxyl group

In bold the correlations deduced from the HMBC

Figure 15.3.S   ¹H-¹³C correlations in HMBC spectrum of unknown compound **15**.

Finally, after the assignment of the NMR signals, a more detailed explanation of the main mass fragmentations is displayed in Figure 15.4.S.

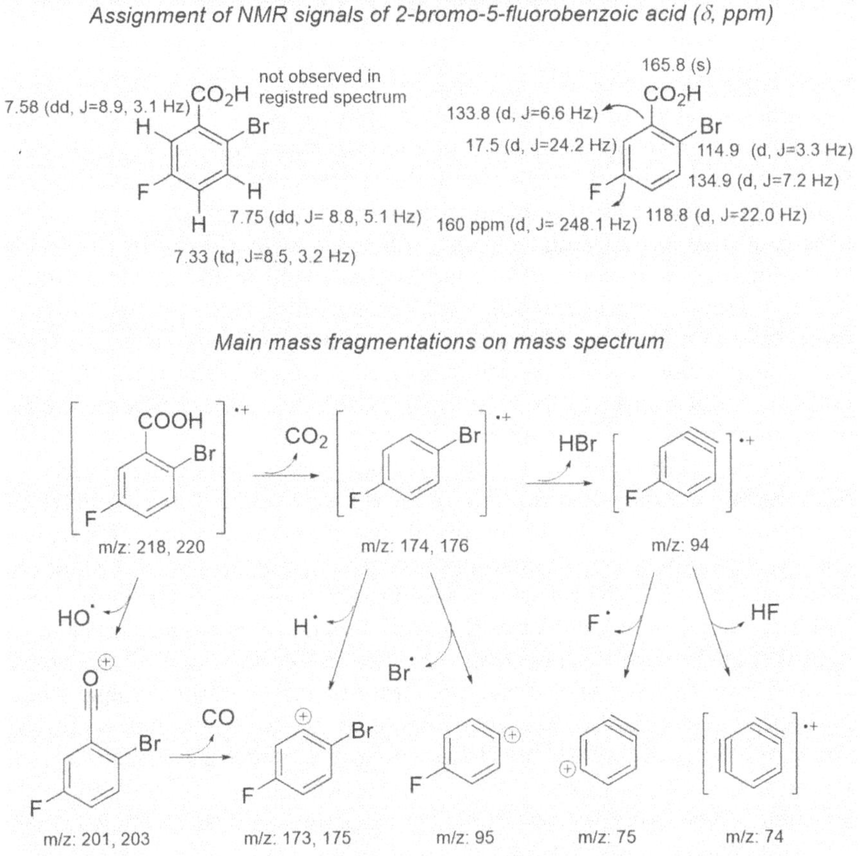

Assignment of NMR signals of 2-bromo-5-fluorobenzoic acid (δ, ppm)

Main mass fragmentations on mass spectrum

Figure 15.4.S NMR signal assignment and mass fragmentations of 2-bromo-5-fluorobenzoic acid.

Problem 16

The mass spectrum shows an odd m/z value (m/z 193) for the molecular ion, which is also the base peak of the spectrum. M-1, M-26, and M-28 fragments are the only additional noteworthy peaks of the spectrum. It looks like a particularly stable nitrogen aromatic compound that suffers little fragmentation. Using the rule of thirteen, $C_{13}H_{23}N$ can be proposed as the molecular formula after inserting nitrogen in the initial formula ($C_{14}H_{25}$). However, the number of hydrogen atoms is too high for an aromatic compound, and the calculation of the double bond equivalents does not provide a whole number, so that formula is not right. A more probable option is $C_{14}H_{11}N$, a structure with ten unsaturations, which make us think of a bicyclic or tricyclic compound.

The ^{13}C NMR shows only seven signals, which is half of the carbons of the molecule and, therefore, there must be an element of symmetry in the structure that obviously divides it into two equal halves. Regarding the chemical shifts, all the carbons are pretty deshielded and resonate between 119 and 149.5 ppm, as expected for an aromatic compound. From DEPT 135 experiment, it is clear that only two carbons are not attached to hydrogen, the one at 129.0 ppm and the most deshielded of all (149.5 ppm), evidently because of the electron-withdrawing effect of the nitrogen atom. No bands corresponding to aliphatic moieties are observed in the infrared spectrum, with all the C-H bond stretching bands above 3000 cm^{-1} (ν_{Csp^2-H} vibrations at 3021, 3045, and 3080 cm^{-1}). Absorptions at 1577 and 1610 cm^{-1} are interpreted as corresponding to $\nu_{C_{ar}-C_{ar}}$ of an aromatic compound.

With regard to the ^1H NMR, five two-proton signals and only one one-proton signal are observed, the latter being a broad singlet at 6.91 ppm, which looks like the resonance of an aniline-type NH proton. In fact, the presence of such a N-H bond would explain the strong and sharp band at 3359 cm^{-1} (ν_{N-H}) in the IR spectrum. The other signals are a singlet at 6.06 ppm, a doublet of doublets at 6.61 ppm (J = 7.9, 1.0 Hz), a triplet of doublets at 6.68 ppm (J = 7.4, 1.2 Hz), a doublet of doublets at 6.73 ppm (J = 7.5, 1.7 Hz) and, finally, a doublet of double doublets at 6.95 ppm (J = 7.9, 7.4, 1.7 Hz). The analysis of the multiplets points to two identical *ortho* disubstituted aromatic rings which would be bound by the two substituents - the NH and the two equivalent alkene CH at 6.06 ppm which show no coupling. This reasoning leads us to propose dibenzoazepine as the most plausible structure.

Thanks to the HSQC, the proton and carbon signals can be correlated, thus confirming that the proton that resonates at 6.91 ppm is bound to a heteroatom (Fig. 16.1.S). The COSY experiment not only supports the already deduced spin system, but also assigns the proton signals. Indeed, COSY provides an unexpected information in this case, the correlation of the chemically equivalent protons at 6.61 ppm and the NH proton that exhibited no noticeable coupling in the ^{1}H NMR spectrum, as shown in Figure 16.1.S.

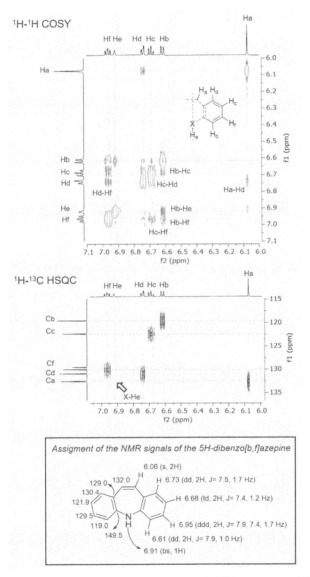

Figure 16.1.S Correlations found in HSQC and COSY spectra and NMR signal assignment of 5*H*-dibenzo[b,f]azepine.

In order to explain the peaks at M-2, M-26, and M-28, although alternative fragmentations (e.g., loss of HCNH˙) can be proposed, we opted for consecutive loss of acetylene and molecular hydrogen (see Fig. 16.2.S for more details).

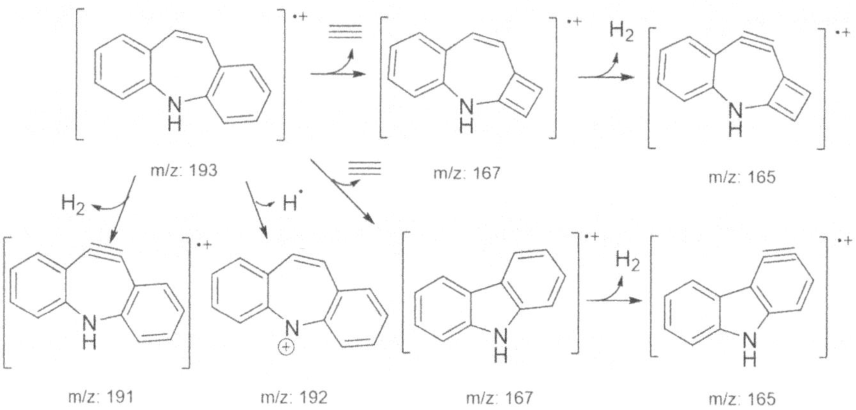

Figure 16.2.S   Fragmentation pattern of 5*H*-dibenzo[b,f]azepine.

Problem 17

The molecular ion at m/z 158 is the base peak of the spectrum (MS), together with the fragment at 129 (M-29). In spite of the presence of abundant peaks corresponding to losses of 15 (m/z 143), 29 (m/z 129), and 43 mass units (m/z 115), the high stability of the molecular ion and the presence of M-28 suggest that we are not dealing with a long linear compound. In addition, peaks such as m/z 77 and 51 are typical in benzenoid compounds. The calculation of the molecular formula from the molecular mass using the rule of thirteen initially provides a hydrocarbon with formula $C_{12}H_{14}$, whose degree of unsaturation is 6.

The $^{13}C\{^1H\}$ NMR shows four slightly deshielded aliphatic carbons between 22.3 and 27.5 ppm and six resonances further downfield, in the region of sp^2 carbons (124.9-142.8 ppm). Since the number of signals is less than the carbons of the molecule, we know that there has to be some local symmetry. The type of carbons must be deduced from the study of ^1H NMR and HSQC experiments since DEPT is not available. Anyway, some information can be obtained from the relative intensity of the signals. We have already mentioned that the area under the signal in broadband proton-decoupled ^{13}C NMR is not proportional to the number of carbons giving rise to it because the heteronuclear Overhausser effect (NOE) from proton decoupling is not equal for all the carbons. Due to this fact and their long relaxation times, a characteristic feature of the signals from quaternary carbons is their low intensity. So, it is easy to guess that the two most deshielded signals, those at 136.7 and 142 ppm, are the resonances of carbons not bound to hydrogen. On the other hand, while being aware that intensity is not conclusive, one can suspect that each of the most intense signals (125.1 and 128.3 ppm) are due to the resonance of two chemically equivalent carbons. Taking into account that there are six unsaturations in the molecule, the hypothesis of a monosubstituted benzene ring should be seriously considered.

The ^1H NMR spectrum shows four two-proton multiplets whose complexity dismisses the idea of a four-atom open-chain. With regards to the chemical shifts, two of the multiplets appear a little more deshielded, approximately centered at 2.27 and 2.47 ppm, suggesting that they are bound to the Csp^2 carbons of an alkene or an aromatic ring. In view of the absence of any large J_{cis} or J_{trans} coupling constants, the one-proton triplet of triplets at 6.18 ppm (J = 3.9, 1.7 Hz) is likely the resonance of a trisubstituted alkene proton. Moreover, the coupling pattern in the aromatic region clearly shows that there is a phenyl ring in the molecule. The monosubstituted benzene is an AA'BB'C spin system with the signals not well resolved and yet, we can easily note a one-proton triplet of triplets-like signal at 7.28 ppm with large (*ortho*) and little (*meta*) coupling constants, a two-proton apparent triplet at 7.36 ppm with a large split and, finally, a signal at approximately 7.43 ppm that shows a large coupling constant. The chemical shifts of the aromatic protons are quite similar, thus indicating that the substituent of the ring does not have a strong electronic effect.

The HSQC spectrum confirms the four aliphatic methylene carbons (signals at 22.3, 23.2, 26.0, and 27.2 ppm) by correlating them to the corresponding two-proton multiplets of the ^{1}H NMR. As proposed, the two most deshielded carbons (δ_C 136.7 and 142.0 ppm) are not attached to hydrogen and therefore, they are the *ipso* carbon of the phenyl ring and the fully substituted alkenyl carbon. The assignment of the hydrogen-bound carbons of the benzene ring and the alkene is straightforward.

The structures that fit the data displayed so far are (cyclopentylidenemethyl) benzene and 1-phenyl-1-cyclohexene (Fig. 17.1.S). The chemical shifts of the exocyclic alkene carbons of the cyclopentylidene derivative are expected to be a little higher. Moreover, the observed signal for the alkene proton is a triplet of triplets with quite different coupling constants, 3.9 and 1.7 Hz. Both allylic couplings would be similar in the case of the five-membered ring and not as high as 3.9 Hz, which is a value usually observed for vicinal couplings between alkene and the neighboring methylene protons (Reich 2018).

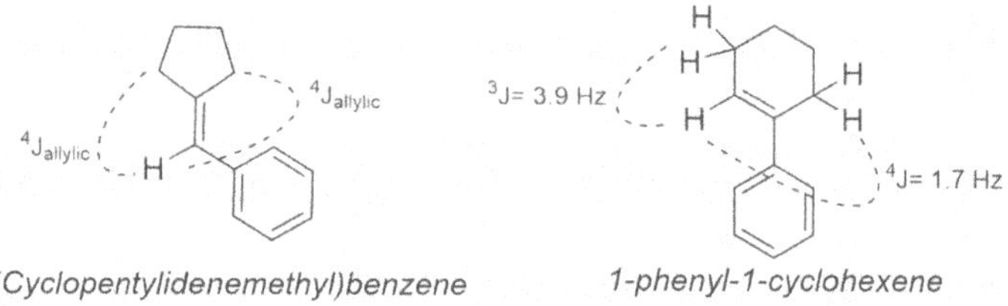

(Cyclopentylidenemethyl)benzene   1-phenyl-1-cyclohexene

Figure 17.1.S   Possible structures for unknown compound **17**.

The ^{1}H-^{1}H COSY experiment gives more information about the aliphatic spin system (Fig. 17.2.S). It is easy to put in order the four methylenes, although a small coupling is visible between the methylene ends. This coupling would be a ^{4}J in the cyclopentylidene derivative and a homoallylic coupling (^{5}J) in the case of the cyclohexene. However, it should be taken into account that the homoallylic coupling can be relatively large when the CH bonds are aligned with the π-orbital of an intervening double bond, as in the coupling between the pseudoaxial protons in the stable half-chair conformation of the cyclohexene derivative (see Fig. 17.2.S). In fact, the homoallylic coupling constants in these structures are usually between 2 and 5 Hz.

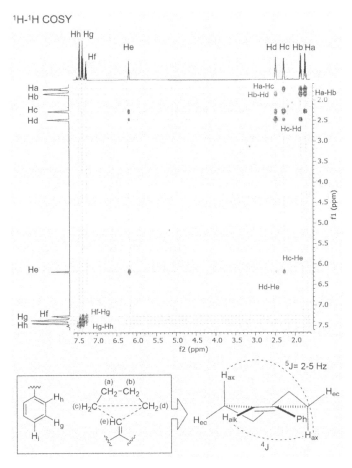

Figure 17.2.S ^{1}H-^{1}H correlations and structural possibilities for unknown compound **17**.

In order to be sure about the right isomer, the results from ^{1}H-^{13}C HMBC experiment should be analyzed. The quaternary alkene carbon shows no correlation with the alkenyl proton or with any of the ring methylenes. The absence of correlation to the neighboring proton is understandable since ^{2}J is usually smaller than ^{3}J and it is often lost in aromatic rings and other conjugated systems (look at the correlations of aromatic carbons in this same HMBC). However, the coupling with the two internal methylenes should be analogous in the case of the five-membered cycle. It should be the same for the alkene CH; the correlations between this carbon and the two methylene-end protons or the two internal methylene protons would not be any different in the case of the exocyclic alkene (Fig. 17.3.S).

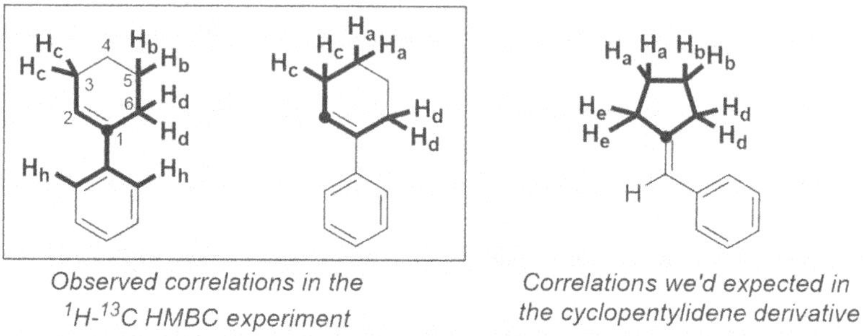

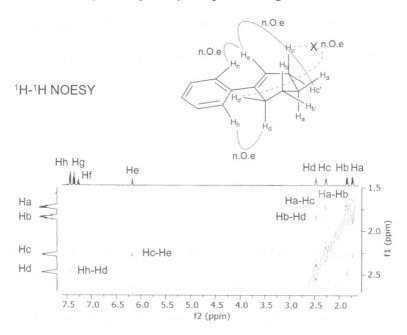

Observed correlations in the
^1H-^{13}C HMBC experiment

Correlations we'd expected in
the cyclopentylidene derivative

Figure 17.3.S ^1H-^{13}C correlations in ^1H-^{13}C HMBC spectrum of unknown compound **17**.

Thanks to the ^1H-^1H NOESY spectrum, we can confirm the assignation of the aliphatic spin system since the *ortho*-protons of the phenyl ring correlate with neighboring methylene protons (Fig. 17.4.S).

Figure 17.4.S ^1H-^1H spatial correlations (NOESY) of unknown compound **17**.

In addition to the NMR signal assignment (Fig. 17.5.S), a graphical explanation of the main fragmentation patterns observed for 1-phenyl-1-cyclohexene is shown in Figure 17.6.S. Along with the loss of acetylene by, for example, a retro Diels-Alder cleavage, and of other neutral molecules, such as ethylene or butene, phenyl-, and other alkyl cleavages give rise to the main fragment ions found in the EI-MS spectrum.

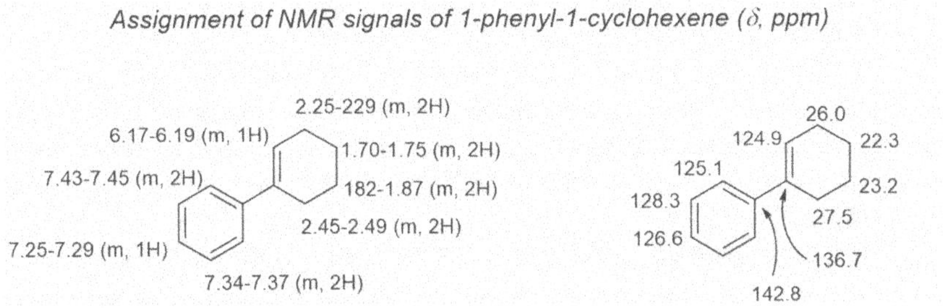

Assignment of NMR signals of 1-phenyl-1-cyclohexene (δ, ppm)

2.25-229 (m, 2H)
6.17-6.19 (m, 1H)
1.70-1.75 (m, 2H)
7.43-7.45 (m, 2H)
182-1.87 (m, 2H)
2.45-2.49 (m, 2H)
7.25-7.29 (m, 1H)
7.34-7.37 (m, 2H)

26.0
124.9
22.3
125.1
23.2
128.3
27.5
126.6
136.7
142.8

Figure 17.5.S   NMR signal assignment for 1-phenyl-1-cyclohexene.

m/z: 158
m/z: 104
m/z: 77

CH$_3$CH$_2$
CH$_3$

m/z: 130
m/z: 129
m/z: 143
m/z: 115
m/z: 51

H

m/z: 129
m/z: 77
m/z: 51

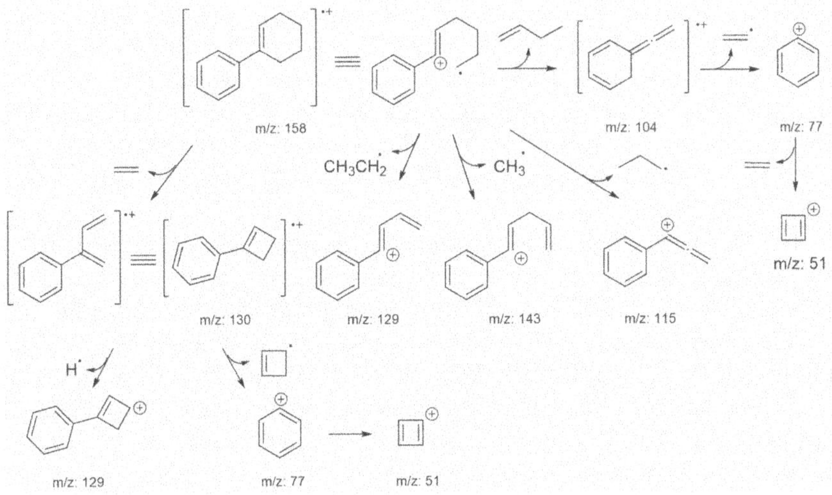

Figure 17.6.S   Main fragmentation patterns of 1-phenyl-1-cyclohexene.

Problem 18

The molecular ion appears as a prominent peak at m/z 189, 18 mass units away from the following fragment (MS). Therefore a nitrogen compound with an OH group should be a promising candidate. In fact, the base peak showing up at m/z 130 (M-59) could correspond to a loss of a carboxymethyl unit, and loss of HCOOH (or CO_2 and H_2) would provide M-46, that is, m/z 143 (9%). However, there is no evidence of the presence of an amine group.

A very intense absorption (IR) at 1683 cm^{-1} ($v_{C=O}$) suggests the presence of a carbonyl or carboxyl group in the molecule. This wavenumber makes us think of a conjugated ketone, an amide, or a carboxylic acid. The quite sharp band at 3396 cm^{-1} likely corresponds to the stretching of the NH bond of a secondary amine, overlapped with a very wide band from 2500 cm^{-1} to 3300 cm^{-1}, which clearly implies the stretching vibration of an OH group of a carboxylic acid. A first attempt at the rule of thirteen would provide $C_{12}H_{15}NO$, an incorrect formula, because the unsaturation index is not an integer. However, a probable molecular formula is obtained by inserting a second oxygen, $C_{11}H_{11}NO_2$, so that the unsaturation index would be 7.

The ^{13}C NMR spectrum supports the presence of the carboxylic carbon on account of the resonance at 174.3 ppm. Upfield of it, seven signals (111.4-136.3 ppm), four CH and three quaternary carbons, are found in the aromatic/olefinic region. The quaternary carbon that resonates at 136.3 ppm is probably the one deshielded by the nitrogen of the molecule. The compound has an aliphatic part consisting of two methylenes (DEPT) appearing at 20.4 and 34.7 ppm. Besides, the slight deshielding of the latter suggests that it could be next to a sp^2 carbon. Only 10 signals are observed and we are considering an 11-carbon molecule, but the intensity of the signal at 118.2 ppm might be indicative of a two-carbon resonance.

The ^1H NMR only shows the signals of 10 protons. The two triplets at 2.59 and 2.94 ppm with typical vicinal couplings in a freely rotating alkyl chain (J = 7.6 ppm) are obviously the resonances of the two methylene units previously observed by ^{13}C NMR. Further downfield, the signals of 5 aromatic protons are found. There are three one-proton doublets at 7.52 (J = 7.8 Hz), 7.54 (J = 8.2 Hz), and 7.11 ppm (J = 2.1 Hz), and, approximately at

6.97 and 7.07 ppm, two apparent triplet of doublets with one large and one small split. The presence of these multiplets is a strong evidence that there is an *ortho*-disubstituted benzene ring in the structure. The doublet at 7.11 ppm, the one with the small coupling constant (2.1 Hz), should not be part of the spin system or, at least, the signals are not sufficiently resolved to assure it. It cannot be a geminal alkene either since the hypothetical coupled alkene proton does not show in the ^{1}H NMR nor the alkene methylene carbon in the DEPT 135 spectrum. Finally, the only one signal left is a very strongly deshielded proton at 10.77 ppm. A signal with such a high chemical shift would usually be assigned to the carboxylic proton, but the expected secondary amine is missing, and the spectrum has been registered in deuterated dimethyl sulfoxide, a solvent that causes large downfield shifts... So, after examining the ^{1}H NMR spectrum, two spin systems are clear, but we still need to put the pieces together. There is a missing proton and the origin of the resonance at 10.8 ppm should also be clarified (Fig. 18.1.S). Moreover, there is another missing unsaturation to be taken into account.

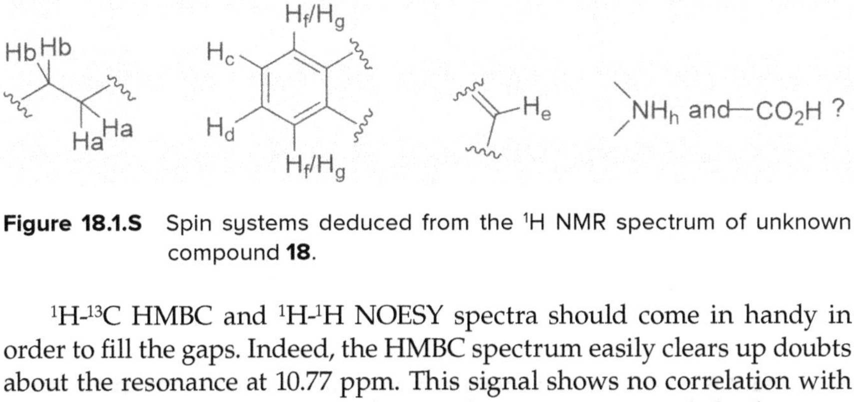

Figure 18.1.S Spin systems deduced from the ^{1}H NMR spectrum of unknown compound **18**.

^{1}H-^{13}C HMBC and ^{1}H-^{1}H NOESY spectra should come in handy in order to fill the gaps. Indeed, the HMBC spectrum easily clears up doubts about the resonance at 10.77 ppm. This signal shows no correlation with the carboxylic carbon, but it does with some protons of the benzene ring and with the proton at 7.11 ppm, thus indicating that the signal at 10.77 ppm corresponds to a secondary amine N-H. Combining the high chemical shift of this NH proton, the discovered connections, and the fact that there is a seventh unsaturation in the molecule, one realizes that the unknown compound is an indole derivative. Accordingly, the split of the signal at 7.11 ppm is due to the ^{1}H-^{1}H coupling with the NH. On the other hand, the carboxylic carbon is coupled with the alkyl methylenes and, therefore, it is in the alkyl chain. From the correlations of the quaternary carbons and aromatic protons, it is possible to unequivocally locate this alkyl substituent at the C3 of the indole ring, as well as to assign some of the aromatic proton signals (Fig. 18.2.S).

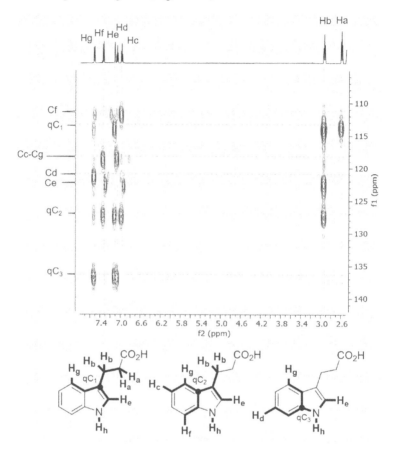

Figure 18.2.S   Some ¹H-¹³C correlations deduced from the HMBC spectrum of unknown compound **18**.

All the expected spatial correlations are found in the ¹H-¹H NOESY spectrum, thus facilitating a full assignment of the aromatic protons (Fig. 18.3.S).

On account of the above data, the NMR signal assignment of this degradation product from tryptophan is displayed in Figure 18.4.S.

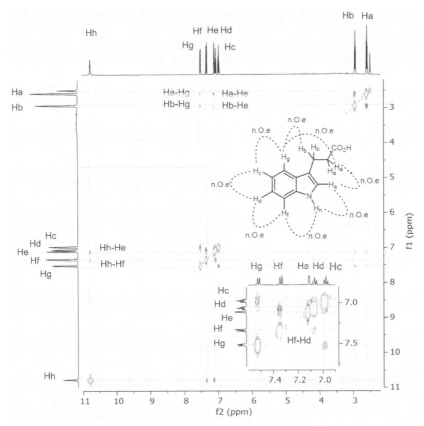

Figure 18.3.S Through space ^1H-^1H correlations in NOESY spectrum of 3-(1H-indol-3-yl)propanoic acid.

NMR Signal Assignment (δ, ppm)

2.94 (t, J=7.6 Hz, 2H) CO$_2$H
2.59 (t, J=7.6 Hz, 2H)
7.52 (d, J= 7.8 Hz, 1H)
6.93-7.00 (m, 1H)
7.05-7.08 (m, 1H)
7.11 (d, J= 2.1 Hz)
7.34 (d, J= 8.2 Hz)
10.77 (s, 1H)

174.3 CO$_2$H
20.4   34.7
118.2
121.0
118.2   122.2
114.4

Figure 18.4.S NMR signal assignment of 3-(1H-indol-3-yl)propanoic acid.

Finally, the mass spectrum shows some typical fragmentations of the indole ring as well as those derived by the presence of the carboxypropyl substituent at C3. Indeed, after decarboxylation or loss of other neutral molecules (H$_2$, HCOOH, H$_2$C = CH$_2$), allylic/benzylic cleavages generate the base peak at m/z 130. Other characteristic fragmentations are the loss HCN and HCNH (Fig. 18.5.S).

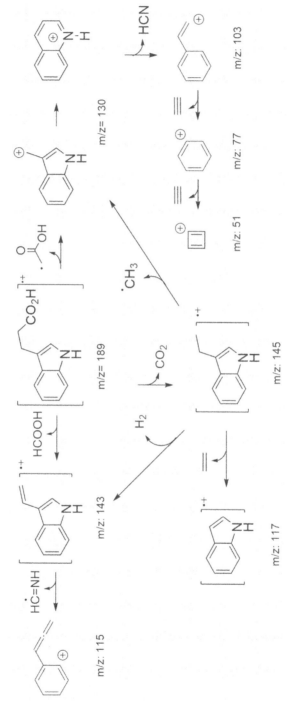

Figure 18.5.S Main mass fragmentations of 3-(1*H*-indol-3-yl)propanoic acid.

Problem 19

A glance at the mass spectrum reveals that the molecular ion (m/z = 94) easily loses ethylene to provide the base peak at m/z 66 (M-28). Another recognizable fragment corresponds to a loss of methyl radical (m/z 79, M-15), thus suggesting that the molecule is a hydrocarbon, or at least it has an alkyl chain.

The ^{13}C NMR spectrum shows only 4 signals, three of them being aliphatic carbons that resonate at 24.7, 41.9, and 48.7 ppm. This last chemical shift could be a hint for the presence of a heteroatom or a strongly deshielding group in its environment. The fourth signal is further downfield and appears at 135.5 ppm, making clear that it is the resonance of a Csp^2 carbon. There is no other signal in this region, so we have to consider that the molecule is symmetric and that we have a symmetric alkene, or a highly symmetric aromatic ring.

The ^1H NMR points out that there are 10 protons in the molecule (or a multiple of 10), which is information that can be used together with the molecular mass to obtain the molecular formula. The compound is consequently a hydrocarbon with formula C_7H_{10} and an unsaturation index of 3. As mentioned before, it must be a disubstituted symmetric alkene, and considering the number of unsaturations, probably a cyclic compound.

The ^1H NMR spectrum shows six multiplets, some of which are quite complex. Due to the fact that there are only four non-equivalent carbons (^{13}C NMR), it is easy to deduce that the structure is such that the protons of several carbons are diastereotopic. The alkene protons are obviously the most deshielded, and resonate at 6.00 ppm, giving rise to a triplet with a small coupling constant (J = 1.7 Hz). The next signal upfield is the two-proton multiplet centered at 2.85 ppm, which would correspond to two protons deshielded by the double bond, although the chemical shift is high enough to question if the anisotropy of the C=C bond is the only effect to be accounted for. In addition, two two-proton multiplets centered at aproximately 0.97 and 1.62 ppm and two one-proton signals, one doublet at 1.09 (J = 8.0 Hz), and one multiplet between 1.31 and 1.34 ppm can be observed. Taking into account the observed integration, these protons are probably located in a plane of symmetry that divides the molecule in half, generating four pairs of chemically equivalent protons.

The ^{1}H-^{13}C HSQC confirms that the multiplets at 0.95 and 1.60 ppm are the signals of two pairs of diastereotopic protons bound to the two equivalent carbons that resonate at 24.7 ppm. The protons at 1.09 and 1.33 ppm correlate to the carbon at 48.7 ppm and, therefore, are also diastereotopic (Fig. 19.1.S.). Once the three methylenes are confirmed, each of the two-proton signals at 6.00 and 2.8 ppm must correspond to the resonances of two equivalent CH units - the alkene protons and the neighboring aliphatic CH protons. The only way to put the pieces together is if the compound is norbornene, that is, bicyclo[2.2.1]hept-2-ene. This constrained structure would also explain the unusually high chemical shift of the bridgehead CH protons.

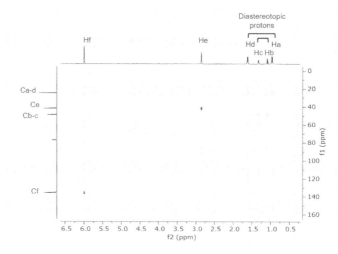

Figure 19.1.S ^{1}H-^{13}C correlations in HSQC of unknown compound **19**.

The multiplets of the ^{1}H NMR are difficult to analyze, so in this case the ^{1}H-^{1}H COSY experiment is helpful to analyze the couplings in the molecule and confirm what we already know (Fig. 19.2.S). The analysis of the spin systems is usually complex in strained cyclic systems without a knowledge of the structure. Indeed, coupling constants between neighboring atoms can even be null depending on the dihedral angle, and long distance couplings become common if the bonds are arranged in the proper manner, as occurs in the W-couplings. In this case, we can see that the alkene proton (Hf) correlates with the neighboring proton He and with one of the protons of the methylene bridge, specifically with the one that resonates at 1.09 ppm (Hb). Similarly, the other proton in the methylene bridge (Hc) correlates with the proton at 0.93 ppm (Ha). Interestingly, the proton in the bridgehead CH (He) shows a faint coupling with this proton

Ha and, however, is strongly coupled to the other diastereotopic proton (Hd).

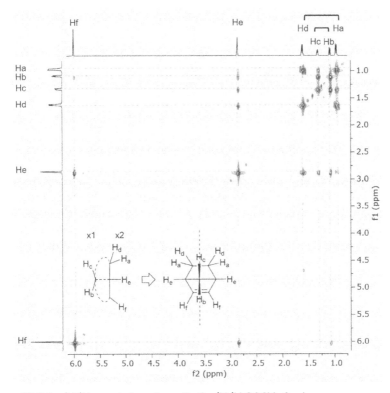

Figure 19.2.S ^{1}H-^{1}H correlations found in the ^{1}H-^{1}H COSY of unknown compound **19**.

The couplings observed in the ^{1}H-^{1}H COSY experiments are in agreement with the reported coupling constants in the norbornene derivatives (Davis and Van Auken 1965). It is noteworthy that the proton of the methylene bridge located above the plane of the double bond is downfield shifted in spite of being in the shielding region of the double bond (Fig. 19.3.S). This exception has been reported by Tori et al. (Tori et al. 1964) and has been supported by *ab initio* MO chemical shift calculations (Martin et al. 1998).

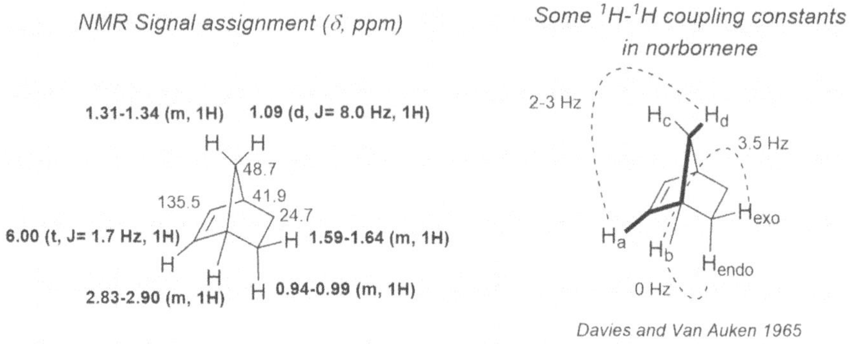

NMR Signal assignment (δ, ppm)   *Some ^{1}H-^{1}H coupling constants*
in norbornene

1.31-1.34 (m, 1H)   1.09 (d, J= 8.0 Hz, 1H)

6.00 (t, J= 1.7 Hz, 1H)

2.83-2.90 (m, 1H)   H 0.94-0.99 (m, 1H)

H 1.59-1.64 (m, 1H)

2-3 Hz

3.5 Hz

H_{exo}

H_{endo}

0 Hz

Davies and Van Auken 1965

Figure 19.3.S   NMR signal assigment of norbornene.

Alkyl rearrangements affect the fragmentation pattern of norbornene and many other terpenes. As shown in Figure 19.4.S, a rearrangement facilitates the loss of methyl radical (M-15) by an allylic cleavage. Retro Diels-Alder processes followed by loss of hydrogen atom generate the rest of the peaks of the MS spectrum, including the base peak at m/z 66.

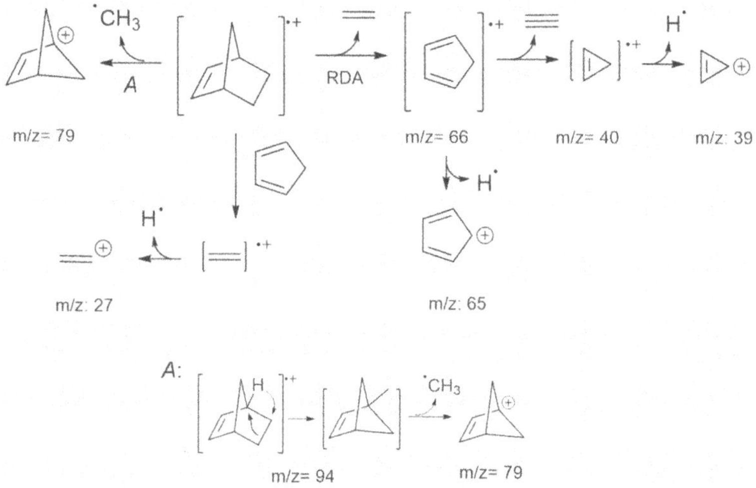

Figure 19.4.S   Main fragmentations of norbornene (MS).

Problem 20

The molecule is likely an alcohol due to the fact that in the mass spectrum, the molecular ion ($m/z = 134$) loses water easily ($m/z = 116$). The relative abundance of the ions generated by the loss of hydrogen atom from these two ion radicals (peaks at m/z 133 and m/z 115) indicates that in both, the positive charge is well stabilized, thus suggesting a compound with benzyl or allylic positions. The loss of 28 mass units which could correspond to an ethylene fragment or carbon monoxide are also recognizable.

The infrared spectrum confirms the presence of a hydroxyl group showing a broad band centered at 3339 cm^{-1}. On the other hand, although weak, the absorptions at 1476 and 1606 cm^{-1}, as well as the stretching of CH bonds above 3000 cm^{-1} (3070, 3023 cm^{-1}), could be indicative of an aromatic ring. In addition, v_{Csp^3-H} stretching bands at 2940 and 2885 cm^{-1} support the idea of an aliphatic moiety in the molecule.

Using the mass of the molecular ion and considering the presence of oxygen, we obtain $C_9H_{10}O$ as the possible molecular formula of the compound, which would have 5 double bond equivalents. Indeed, the ^{13}C NMR shows that there are nine non-equivalent carbons in the molecule. An aliphatic part consisting of two methylene carbons resonating at 29.7 and 35.7 ppm, and a methine unit at 76.2 ppm, obviously downfield shifted by an oxygen atom, can be distinguished. The absence of other carbon resonances in this region leads to discard the idea of an alkyne since the molecule is not symmetric and there are no chemically equivalent carbons. The six peaks between 124 and 145 ppm, two of them (143.2, 144.9 ppm) not bound to hydrogen (DEPT), indicate the presence of a disubstituted benzene ring in the molecule. Taking into account the connectivity of the hydrocarbon skeleton, a logical structure in agreement with the unsaturation index should be cyclic. Accordingly, it is an indanol, and to be precise, 1-indanol, because the compound is not symmetric.

The ^{1}H NMR shows the resonances of the four aromatic protons, a one-proton multiplet, and a three-proton multiplet between 7.23 and 7.42 ppm, and the four signals due to the protons of the aliphatic spin system; a quartet at 5.22 ppm ($J = 5.9$ Hz), two doublets of double doublets at 3.05 ppm ($J = 15.9, 8.5, 4.8$ Hz) and 2.81 ppm ($J = 15.7, 8.2, 6.7$ Hz), and two doublets of double doublet of doublets at 2.47 ($J = 13.1, 8.2, 6.8, 4.8$ Hz) and 1.94 ppm ($J = 13.3, 8.5, 6.7, 5.4$ Hz). The highly deshielded quartet must be the resonance of the hydrogen on the carbon next to the hydroxyl group, being the observed multiplicity due to the accidental equivalence of three vicinal couplings, one with the OH proton and two with the protons of the neighboring methylene. In fact, we can see that the OH signal this time is quite sharp, showing the coupling with the carbinol proton (doublet at 2.2 ppm with J = 5.8 Hz).

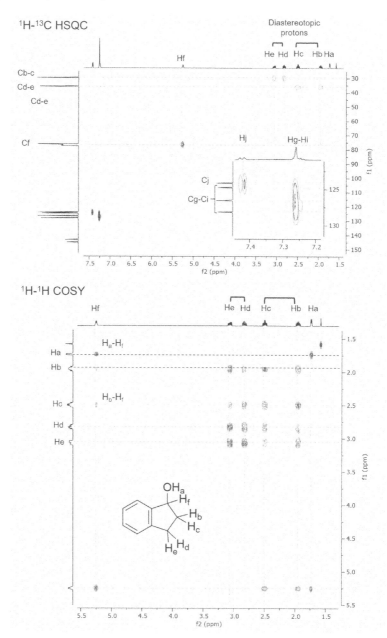

Figure 20.1.S ^{1}H-^{13}C and ^{1}H-^{1}H correlations in 1-indanol.

The complexity of the aliphatic signals is consistent with the presence of a rigid system with diverse spin couplings. Unlike what happens in cyclohexane rings, the vicinal couplings in 5-membered rings are very variable. For cyclopentane type compounds in envelope conformations $J_{cis} > J_{trans}$ in the flat part of the envelope, whereas in twist conformations, the trend is the opposite (Reich 2018). In most cyclopentanes cis protons usually have dihedral angles close to 0°, and trans near 120°, thus making cis couplings usually larger than the trans ones. However, in other cases, cis and trans couplings are quite similar. This is the reason why the stereochemistry assignments are hard to do in this type of cycles using the size of couplings alone.

In spite of the complexity, the largest coupling constants in the signals (J ~13.2 and ~15.8 Hz make quite clear which protons are in the same methylene (diastereotopic protons), which is information that is easily confirmed by ^1H-^{13}C HSQC (Fig. 20.1.S). ^1H-^1H COSY spectrum allows us analyze the connectivity of the aliphatic system. As expected, the protons at 3.5 and 2.8 ppm are in the methylene close to the aromatic ring.

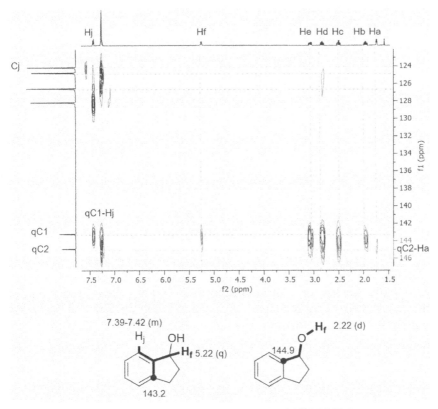

Figure 20.2.S Some key correlations from the ^1H-^{13}C HMBC spectrum.

The overlapping of the signal of three of the aromatic protons hinders a complete assignment of the carbon signals by analysis of the ^1H-^{13}C HMBC experiment. Still, some valuable information can be extracted from it (Fig. 20.2.S). The quaternary carbon at 144.9 ppm correlates to the hydroxyl proton and, therefore, has to be next to the CHOH unit. The other quaternary carbon correlates to the quartet at 5.22 ppm (H_f) and to the aromatic proton at 7.42 ppm (H_j), which is a correlation that is understandable, considering that three-bond couplings are the largest ones in aromatic rings (Fig. 20.2.S).

The assignment of the rest of the carbon signals is difficult with the available data, and NOESY spectrum is not very useful in this case. The signals that can be unequivocally assigned are shown in Figure 20.3.S.

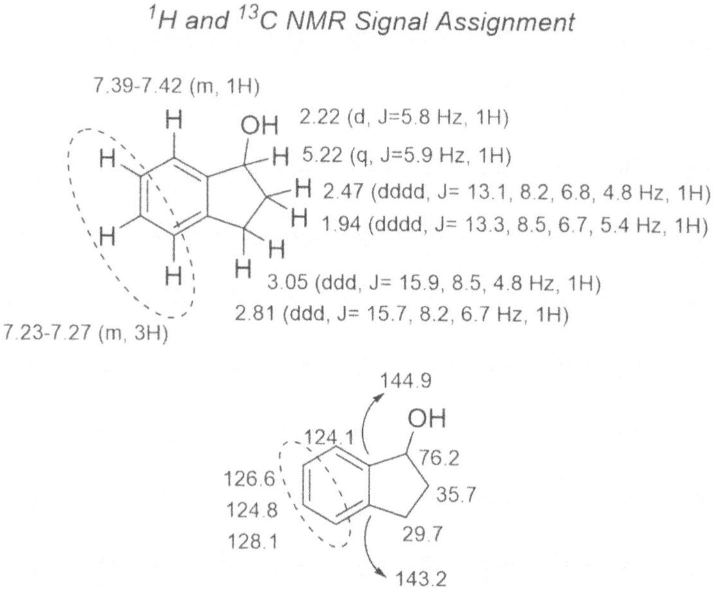

^1H and ^{13}C NMR Signal Assignment

7.39-7.42 (m, 1H)

H   OH   2.22 (d, J=5.8 Hz, 1H)
     H 5.22 (q, J=5.9 Hz, 1H)
     H 2.47 (dddd, J= 13.1, 8.2, 6.8, 4.8 Hz, 1H)
     H 1.94 (dddd, J= 13.3, 8.5, 6.7, 5.4 Hz, 1H)

H   H   3.05 (ddd, J= 15.9, 8.5, 4.8 Hz, 1H)
          2.81 (ddd, J= 15.7, 8.2, 6.7 Hz, 1H)

7.23-7.27 (m, 3H)

144.9
OH
124.1
126.6   76.2
124.8   35.7
128.1   29.7
143.2

Figure 20.3.S   NMR signal assignment of 1-indanol.

Finally, the loss of hydrogen radicals from benzylic positions and of neutral molecules, such as water, ethylene, cyclopropanol, and acetylene explain the main fragmentations in EI-mass spectrum (Fig. 20.4.S).

Main Mass Fragmentations in EI-MS

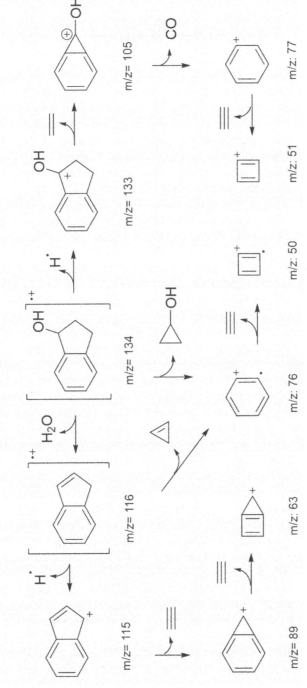

Figure 20.4.S Main mass fragmentations in EI-MS of the 1-indanol.

Problem 21

The molecular ion (m/z 224) is the base peak of the mass spectrum where two abundant peaks corresponding to the loss of aromatic units (m/z 147, M-77 or m/z 120, M-104) and the fragments themselves (m/z 104 and 77) can be easily distinguished. The aromatic ring of the compound is confirmed by the infrared spectrum, which shows the C_{ar}-C_{ar} stretching vibrations at 1605 and 1576 cm^{-1}. Moreover, a strong absorption at 1667 cm^{-1} reveals the presence of a carbonyl group.

The ^{13}C NMR supports this observation and evidences the presence of a ketone by showing a resonance at 192.1 ppm. Ten aromatic carbons make clear that there is more than one benzene ring in the molecule. The strong deshielding of the carbon at 161.7 ppm indicates that there is another oxygen in the molecule. Furthermore, the chemical shift of the carbonyl carbon discards the presence of an ester, so the two oxygen atoms are not in adjacent positions. Two aliphatic carbons are also observed in the spectrum. One of them resonates at 44.8 ppm, probably deshielded by an aromatic ring, and the other one at 79.7, obviously downfield shifted by the second oxygen, so we have an aliphatic-aromatic ether.

Considering the aforementioned two oxygen atoms and the mass of the molecular ion, $C_{15}H_{12}O_2$ is calculated as the molecular formula, so the compound contains a total of 10 unsaturations. The two aromatic rings and the ketone would account for nine, so there is still one missing unsaturation, which is necessarily a cycle, probably a benzene-fused six-membered cycle where we have to locate the oxygen, the ketone, and the two aliphatic carbons.

The presence of the cycle is evident when analyzing the coupling constants of the aliphatic signals in the ^1H NMR spectrum, three one-proton doublets of doublets centered at 2.90, 3.10, and 5.49 ppm. On account of the same large constant due to geminal coupling (17.0 Hz), the former two protons belong to the same methylene unit. Their vicinal coupling constants differ, being much larger for the signal at 3.10 ppm (13.3 Hz). This value is consistent with a ^3J between two axial positions in a cyclohexane-type chair, where dihedral angle is close to 180°. The other axial proton is evidently the one at 5.49 ppm, whose signal shows one large (13.3 Hz) and one small split (2.8 Hz), because of $^3J_{ax-ax}$ and $^3J_{ax-eq}$, respectively.

With regards to the chemical shift, the strong deshielding of the proton at 5.49 ppm is evidence that it is next to the oxygen atom and an aromatic ring. The methylene protons are upfield, but they still suffer the deshielding by the carbonyl group and the weakened effect of the ether oxygen.

On the other hand, the spectrum shows 5 multiplets in the aromatic region. The integration of the signals make clear that some signals of non-equivalent protons are overlapped. The only multiplet that can be readily analyzed is the double doublet at 7.95 ppm, which shows an *ortho* (J = 8.1 Hz) and *meta* (1.8 Hz) coupling. The high chemical shift suggests that either an electron-withdrawing ring substituent decreases the electronic density in that position, or the proton is in a deshielding region due to the magnetic anisotropy of a neighboring group, or both. Therefore, that proton is likely next to the carbonyl carbon.

Besides, although it is a second-order spin system and the corresponding lining effect and distortion of the signals prevent us from doing a full analysis, a triplet of triplets-type signal that can be distinguished at 7.06 ppm probably corresponds to the *"para"* proton of a phenyl group. The resonance of the neighboring protons could be the triplet type signal that integrates for two protons a little downfield.

The ^{1}H-^{13}C HSQC confirms that they are four quaternary carbons, the ketone at 192.1 ppm, the oxygen-bound aromatic carbon at 161.7 ppm, and two more at 121.1 and 138.9 ppm. The spectrum also identifies the diastereotopic protons at 2.90 and 3.10 ppm, whose signals correlate to the carbon at 44.8 ppm. As expected, the multiplet between 7.05 and 7.08 ppm contains the signals of two non-equivalent protons, which correlate to the carbons at 118.3 and 121.8 ppm. Likewise, the multiplet centered at 7.51 ppm is owing to two chemically equivalent aromatic protons that are coupled to the carbons at 126.3 ppm and the proton correlated to the carbon at 136.3 ppm (Fig. 21.1.S).

The ^{1}H-^{1}H COSY spectrum should allow location of the protons in each aromatic ring, but the overlapped signals make it difficult for those of the disubstituted benzene ring. Since the two substituents are connected forming a cycle, it is clear that they must be *ortho*. However, the assignment of the signals is puzzling except for the doublet at 7.95 ppm, which is undoubtedly the signal of one of the protons at the end of the spin system.

The ^{1}H-^{13}C HMBC spectrum confirms that the carbonyl carbon is next to the proton that resonates at 7.95 ppm since this signal correlates strongly to the carbon at 192 ppm. Besides, it facilitates the assignment of the rest of the quaternary carbons (Fig. 21.2.S). The carbon at 138 ppm (qC$_2$) correlates with the two-proton signal at 7.45 ppm (H$_g$), and therefore it is in the phenyl ring. Besides, as usual, while the proton at 7.95 ppm (H$_j$) shows a strong correlation to the quaternary O-carbon (qC$_3$) due to a three-bond coupling, no cross peak due to a two-bond correlation appears in the case of the carbon that resonates at 121.1 ppm (qC$_1$). In this sense, the strong coupling between the carbon at 136 ppm (C$_i$) and the signal at 7.95 ppm ((H$_j$) suggests that they are in relative *meta* positions in the disubstituted benzene ring. This suggestion is supported by the weak homonuclear correlation observed in the ^{1}H-^{1}H COSY spectrum between those protons.

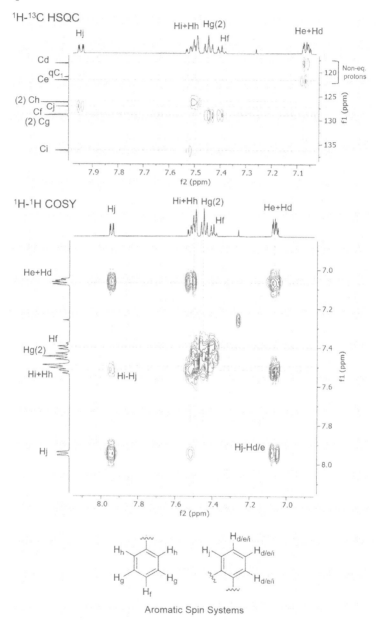

Figure 21.1.S   Analysis of the aromatic spin systems of unknown compound **21**.

Finally, the assignment of the carbons at 118.3 ppm (C_d) and 121.8 ppm (C_e) and the corresponding protons is dubious. The weak correlation of the carbon at 118.3 ppm (C_d) to the protons at 7.95 ppm (H_j) and 7.5 ppm (H_i) might imply that the carbon is two-bond coupled to those protons. However, the overlapping of the proton signals makes it difficult to corroborate this hypothesis.

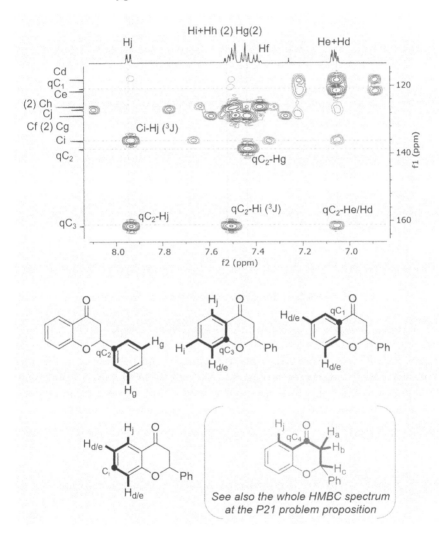

Figure 21.2.S Some ^{1}H-^{13}C correlations found in the HMBC spectrum.

With all this data in hand, an assignment of the NMR signals of the molecule deduced, 2-phenylchroman-4-one (commonly known as flavanone) is depicted in Figure 21.3.S.

NMR Signals Assignment

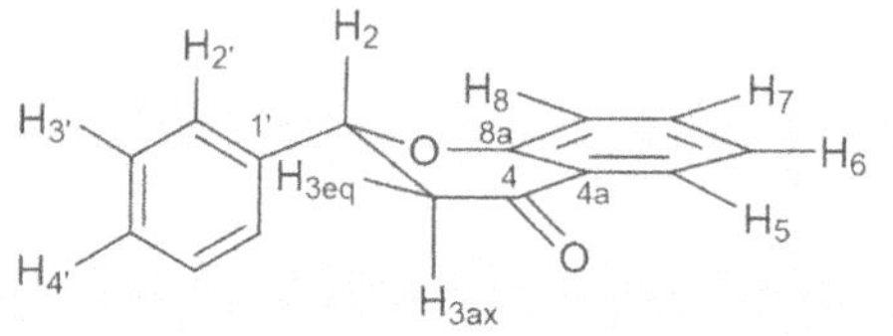

C_2: 79.7	$C_{1'}$: 138.9
C_3: 44.8	$C_{2'}$ and $C_{6'}$: 126.3
C_4: 192.1	$C_{3'}$ and $C_{5'}$: 129.0
C_{4a}: 121.1	$C_{4'}$: 128.9
C_5: 127.2	
C_6: 118.3/121.8	
C_7: 136.3	
C_8: 118.3/121.8	
C_{8a}: 161.7	

H_2 : 5.49 (dd, J= 13.3, 2.8 Hz, 1H)

H_{3eq}: 2.90 (dd, 1H, J=16.9, 2.8 Hz, 1H)

H_{3ax}: 3.10 (dd, J= 17.0, 13.3 Hz, 1H)

$H_{2'}$ and $H_{6'}$: 7.53-7.49 (m, 3H, overlapped with H_7)

$H_{3'}$: 7.48 – 7.42 (m, 2H)

$H_{4'}$: 7.42 – 7.37 (m, 1H)

H_6 and H_8: 7.05-7.08 (m, 2H)

H_7: 7.53-7.49 (m, overlapped with $H_{2'}$ and $H_{6'}$)

H_5: 7.95 (dd, J= 8.1, 1.8 Hz, 1H)

Figure 21.3.S   NMR signal assignment of flavanone.

Finally, a more detailed analysis of the mass spectrum confirms the above structure and reveals the fragmentations displayed in Figure 21.4.S. Retro Diels-Alder cleavages of molecular ion provide m/z 120 and m/z 104. Other fragmentations from the same radical cation involve α-, benzylic, and phenylic cleavages and the loss of other neutral molecules. Decarbonylations and loss of acetylene complete the fragmentation pattern of flavanone.

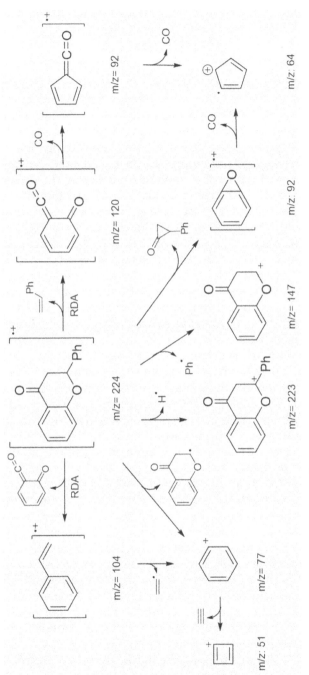

Figure 21.4.S Main mass fragmentations of flavanone.

Problem 22

The molecular ion (m/z 150) readily loses a methyl radical (M-15, m/z 135) and a fragment of 28 mass units (M-28, m/z 107), probably carbon monoxide or ethylene. The absorptions at 1678 cm^{-1} (intense) and 1617 cm^{-1} in the infrared spectrum indicate that both losses are possible, and from the frequency of the C=O stretching, it could be suggested that the carbonyl group is conjugated with a C=C moiety. Considering the presence of an oxygen atom, the rule of thirteen provides a molecular formula of $C_{10}H_{14}O$, which would correspond to a structure with 4 double bond equivalents.

The ^{13}C NMR shows a strongly deshielded carbon at 204.1 ppm, thus confirming the presence of a ketone carbonyl. The appearance of the alkene carbons at 121.3 and 170 ppm is consistent with an enone. Indeed, the difference in chemical shifts between both alkene carbons is probably due to the electron density withdrawal from the carbonyl group on β position. The rest of the carbons appearing below 60 ppm are Csp^3 and, therefore, the unknown compound **22** has to be a bicycle in order to fit the degree of unsaturations.

From the ^1H NMR spectrum, on the basis of the one-proton signal deshielded enough to be an alkene proton (the multiplet at 5.7 ppm), it can be deduced that the alkene is trisubstituted. The chemical shift, which is not too high, indicates that this is not the position deshielded by the conjugation with the ketone, thus suggesting that the carbonyl group is attached to that very alkene carbon. Besides, the multiplicity of the signal makes clear that the proton is coupled to protons at allylic and/ or homoallylic positions. This idea is supported by the small 1 Hz split of the three-proton doublet at 2.0 ppm. Additionally, two three-proton singlets at 0.99 and 1.48 ppm indicate the presence of two more methyl groups without adjacent protons. The remaining four protons give rise to a doublet at 2.06 ppm (J = 9.1 Hz), two triplets at 2.40 ppm (J = 5.8 Hz) and 2.63 ppm (J = 5.9 Hz), and a doublet of triplets at 2.79 ppm (J = 9.2, 5.4 Hz). The diverse coupling constants, which are in agreement with dihedral angles rigidly fixed by the geometry of a bicyclic compound, tells us that they all are non-chemically equivalent and part of the same spin system.

The ^1H-^{13}C HSQC spectrum confirms that the quaternary alkene carbon (170.2 ppm) is at the β position from the carbonyl group and defines two CH groups at 57.7 and 49.8 ppm, whose protons resonate, respectively, at 2.63 and 2.40 ppm, and a CH$_2$ group at 40.9 ppm, being the two corresponding diastereotopic protons - those at 2.40 and 2.63 ppm.

The spectrum also reveals that the carbon at 54.1 ppm is not bound to hydrogen and, therefore, it should be the one bound to the two shielded methyl groups at 0.99 and 1.48 ppm.

The ^{1}H-^{1}H COSY spectrum provides more details on the homonuclear coupling in the molecule. As shown in Figure 22.1.S, the two methyl groups that show no coupling in ^{1}H NMR appear correlated in COSY spectrum. The coupling would be a ^{4}J coupling, which is quite usual in rigid and/or conjugated systems. On the other hand, the alkene proton (H$_h$) correlates not only to the methyl group at 2.00 ppm (H$_c$), as expected, but also to the protons at 2.40 ppm (H$_e$) and 2.63 ppm (H$_f$), probably located at allylic positions. At the same time, these protons are coupled to each other and, interestingly, to only one of the diastereotopic protons of the methylene of the molecule, to be precise, to the proton at 2.79 ppm (H$_g$). Although this situation is unusual, very small ^{3}J values have been reported for polycyclic systems with dihedral angles close to 90°.

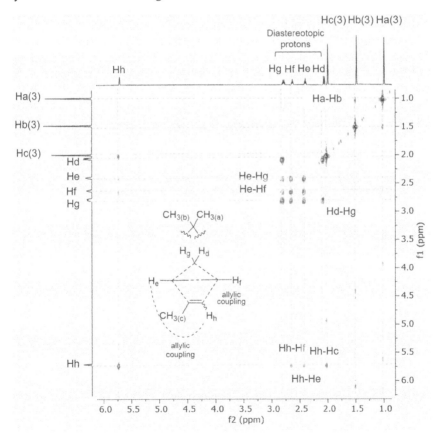

Figure 22.1.S ^{1}H-^{1}H correlations found in the COSY spectrum of unknown compound **22**.

In order to analyze the spin system more closely, the correlations found in COSY spectrum, together with the coupling constants of the signals, must be analyzed in this case. The two triplets with analogous coupling constant (J ~5-6 Hz) at 2.40 ppm (H_e) and 2.63 ppm (H_f) indicate that these methine protons are similarly coupled to the methylene proton at 2.79 ppm (H_g), thus suggesting not only comparable torsional angles for both, but also that the coupling between them has a similar magnitude since the signals are triplets. It may seem that all the couplings are vicinal, but the CMe_2 unit must still be located. It should not be placed adjacent to the C=C double bond or to the ketone carbonyl, because that would imply a strong coupling through 5 bonds, which is something quite unusual (Fig. 22.2.S-A). Taking this into account, different bicyclic systems having all the pieces of the puzzle can be proposed (Fig. 22.2.S-B). The cyclopropane derivatives (IV-V) are readily discarded because much lower chemical shifts for the aliphatic carbons in ^{13}C NMR spectrum are expected in these cases. Among the cyclobutane derivatives (I-III), 4,6,6-trimethylbicyclo[3.1.1]hept-3-en-2-one, commonly named as verbenone (I), is a more probable structure. Indeed, the theoretical ^1H and ^{13}C chemical shifts are in accordance with the observed values, and the spatial arrangement of the protons in this molecule explains the magnitude of the J couplings found in ^1H NMR and ^1H-^1H COSY spectra (Fig. 22.2.S-C). The analogous dihedral angles between the diastereotopic proton H_g and the two bridgehead CH protons (H_e and H_f) are consistent

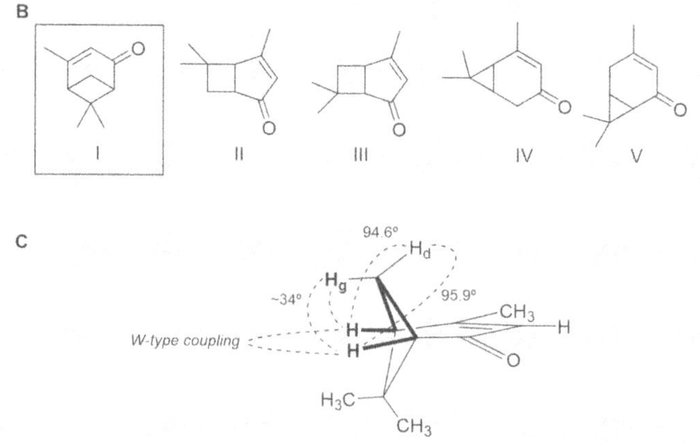

Figure 22.2.S Possible spin systems rationalized for unknown compound **22**.

with the similar J values observed. Besides, the angle of approximately 95° in the case of the other methylene proton (H$_d$) justifies the absence of coupling in this case. Finally, the strong ^{4}J coupling between the CH protons (H$_e$ and H$_f$) is unusual but possible if the disposition of the bonds is adequate (W-arrangement). The overlapping of orbital lobes behind the CH bonds has been used to rationalize this large ^{4}J coupling (5.8-6.4 Hz) in verbenone and analogous compounds, such as α-pinene and myrteral (Huckerby 1970).

The ^{1}H-^{13}C HMBC spectrum can be used to assign the signals of the bridgehead CH protons H$_e$ and H$_f$, thanks to the correlations of the carbonyl and the methyl group on the alkene (Fig. 22.3.S). The spectrum shows the expected cross peaks though, as usual, ^{2}J couplings are missing sometimes. It is noteworthy that the methylene carbon (C$_{dg}$) does not correlate to any proton, probably due to the very small ^{2}J couplings and the absence of protons 3 bonds away.

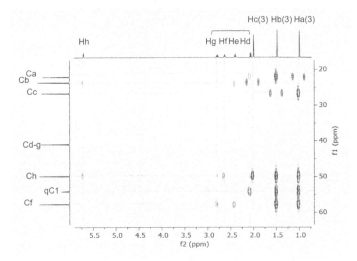

Some ^{1}H-^{13}C Correlations in ^{1}H-^{13}C HMBC Spectrum

Figure 22.3.S Correlations found in the ^{1}H-^{13}C HMBC spectrum of verbenone.

On the other hand, the correlation of the methyl group at 1.48 ppm (H$_b$) to the methylene proton at 2.79 ppm (H$_g$) in the ^{1}H-^{1}H NOESY spectrum allows us to unequivocally assign the signals of the methyl groups of the CMe$_2$ unit (Fig. 22.4.S.). The methyl group oriented towards the enone function is shifted upfield due to the shielding region above (and below) the plane defined by the conjugated double bonds.

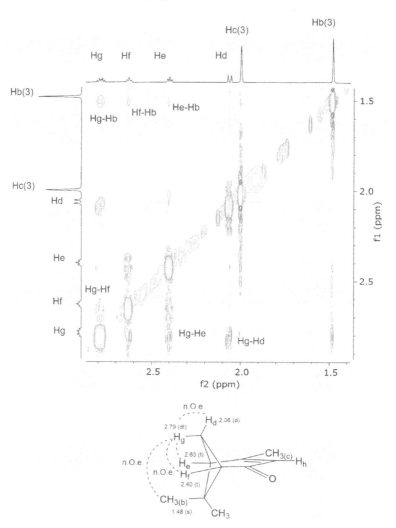

Figure 22.4.S ^{1}H-^{1}H through-space correlations with diastereotopic proton at 2.79 ppm (H$_g$).

With all the collected data, a complete assignment of the NMR signals of verbenone is provided in Figure 22.5.S.

NMR signal Assignment (δ, ppm)

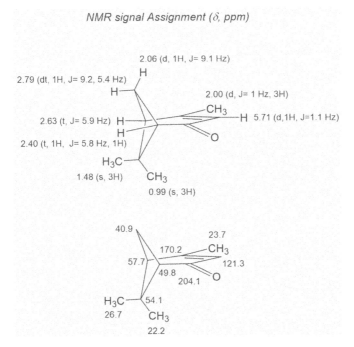

Figure 22.5.S NMR signal assignment of verbenone.

As explained in Problem 19, terpene derivatives are prone to rearrangements under mass ionization conditions. This tendency is clearly observed in the relatively complex fragmentation pattern of verbenone. As shown in Figure 22.6.S, and in addition to the loss of methyl radical and carbon monoxide, the rest of the fragmentations detected are based on several rearrangement pathways (*A-C*) along with the loss of ethylene, acetylene, hydrogen atom, methyl radicals, and further decarbonylations.

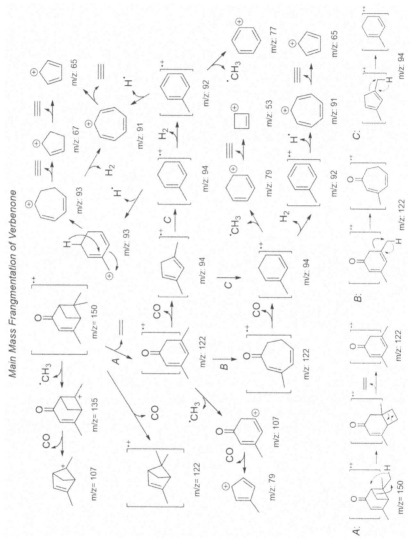

Figure 22.6.S   Fragmentation pattern of verbenone.

Problem 23

The most probable molecular formula obtained from the accurate mass of the molecular ion is $C_{15}H_{12}O_2$ (www.chemcalc.org, Patiny and Borel 2013). Accordingly, the calculation of the degree of unsaturation reveals that the compound contains a total of 10 rings and/or double bonds.

The 15 signals observed in the ^{13}C NMR spectrum indicate that there is no element of symmetry in the molecule. Apart from the resonance at 55.5 ppm, typical chemical shift of a CH_3O group, all resonances correspond to sp² carbons, which is something expected, considering the number of unsaturations of the molecule. The signals at 156.6, 154.0, and 152.3 ppm are probably carbons bound to oxygen atoms. On account of the fact that there are only two oxygen atoms in the molecule, C_{arom}-O-Me and C_{arom}-O-C_{arom} fragments can be predicted rather reliably.

The ^1H NMR spectrum shows the three-proton signal of OMe group at 3.90 ppm and the multiplets of the nine aromatic protons. Except for the signal at 7.50 ppm, the signals are well resolved and can be easily analyzed (Table 23.1.S). From the small coupling constant of the doublet at 7.38 ppm, it can be deduced that there are no protons on the adjacent carbons. However, the presence of two triplets of doublets with a large coupling constant at 7.13 and 7.00 ppm and two doublets of double doublets at 7.24 and 7.19 ppm with two large splits suggests that there are two *ortho*-disubstituted aromatic rings in the molecule. This data and the 7.2 and 7.3-Hz coupling values lead to consider benzofuran, a heteroaromatic ring that contains a C_{arom}-O-C_{arom} fragment and a proton on the furane ring that would show only a long-range coupling if there were a substituent at the adjacent position.

The ^1H-^{13}C HSQC spectrum reveals that among the aromatic signals there are two more quaternary carbons apart from the three C_{arom}-O already noticed, which is something consistent with a monoarylated benzofurane. The ^1H-^1H COSY spectrum clearly defines two aromatic four-spin systems depicted in Figure. 23.1.S. From all the above data, it is straightforward to propose 2-(2-methoxyfenyl) benzofuran as a suitable candidate. Furthermore, the protons of each ring can even be located because only strong couplings are found in this COSY spectrum. Unfortunately, this fact prevents us from observing the correlation due to the long-range homonuclear coupling of the proton at 7.28 ppm (H_g).

TABLE 23.1.S   Analysis of the multiplets in the aromatic region of ¹H NMR spectrum of unknown compound **23**.

δ (ppm)	Multiplet	J (Hz)
6.92	Doublet of doublets	8.3, 1.1
7.00	Triplet of doublets	7.6, 1.1
7.13	Triplet of doublets	7.5, 1.1
7.19	Doublet of doublet of doublets	8.2, 7.2, 1.4
7.24	Doublet of doublet of doublets	8.3, 7.3, 1.7
7.28	Doublet	1.0
7.40-7.47	Multiplet	—
7.51	Doublet of doublet of doublets	7.6, 1.4, 0.7
8.00	Doublet of doublets	7.8, 1.8

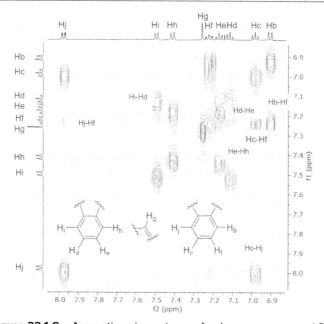

Figure 23.1.S   Aromatic spin systems of unknown compound **23**.

Anyway, thanks to the long-range heteronuclear correlations displayed by the ¹H-¹³C HMBC spectrum, the pieces can be put together (Fig. 23.2.S). The strong correlations of the carbon at 157 ppm with the OMe group and the protons at 6.92 and 8.00 ppm, together with the cross-peaks derived from carbon at 119 ppm, allow for a complete assignment of the spin system of the methoxyphenyl ring. The correlations of the carbons at 130 and 154 ppm do the same with the signals of the benzofuran ring.

The ¹H-¹H NOESY spectrum shows the expected through-space correlations. The coupling between the *ortho* protons in the aromatic spin systems can be observed along with the correlations of the OMe group to the doublet at 7.28 ppm (H$_g$) and to the doublet of doublets at 6.92 ppm (H$_b$)

(Fig. 23.3.S.-A). A full assignment of the NMR signals of the compound is shown below (Fig. 23.3.S.-B).

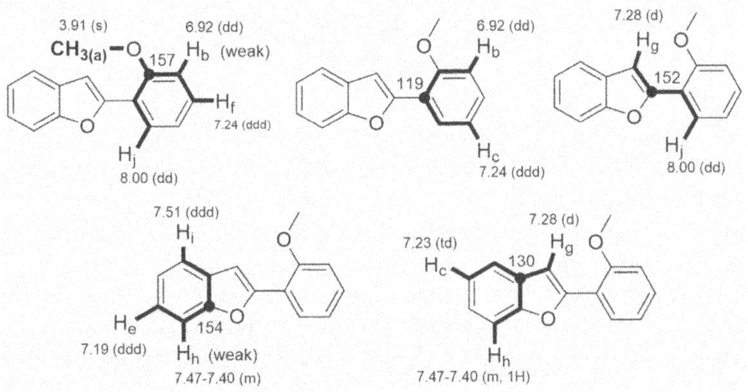

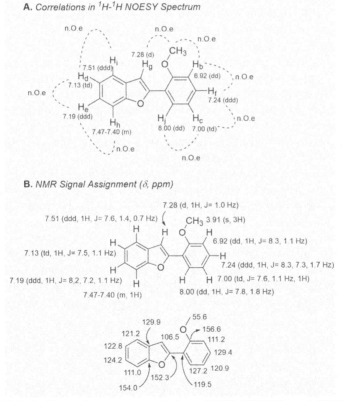

Figure 23.2.S ¹H-¹³C correlations in HMBC spectrum of 2-(2-methoxyphenyl)benzofuran.

Figure 23.3.S Through-space correlations (¹H-¹H NOESY) and NMR signal assignment of 2-(2-methoxyphenyl)benzofuran.

Problem 24

We are probably dealing with a nitrogen compound because the molecular ion shows up at m/z 301 in the EI-MS spectrum. The base peak corresponds to a fragmentation, resulting in the release of a propyl or acetyl unit (m/z 286, M^+-43). This ion subsequently loses 28 mass units, possibly carbon monoxide or ethylene.

The infrared spectrum shows a strong band at 1637 cm^{-1}. This frequency is low for a ketone, even for a conjugated one, but it could be the stretching of a conjugated amide or an acid carbonyl. However, no NH or OH absorption is observed in the spectrum. Considering that it is a nitrogen compound, another option would be the stretch of a C=N bond. On the other hand, the weaker bands at 1570 and 1469 cm^{-1} suggest the presence of an aromatic ring in the structure.

There are only 9 lines in the ^{13}C NMR spectrum. In view of the molecular mass of the compound (301 mass units), the few signals observed indicate that the compound is highly symmetric and/or the heteroatomic fragment is important. We see three aliphatic signals upfield at 18.4, 19.2, and 33.0 ppm, two resonances at 71.1 and 73.0 ppm, which are probably heteroatom-bound aliphatic carbons, and four C_{sp^2} signals, with two of them particularly deshielded (147.0 and 162.3). The chemical shift of the most deshielded carbon is obviously high, but not enough to be an amide carbon, at least not one with a usual chemical shift. A C=N carbon, however, would resonate at such a frequency.

The ^{1}H NMR spectrum reveals that there are 23 protons in the molecule, twenty of which are aliphatic (Table 24.1.S). The two aromatic signals, a one-proton triplet at 7.84 (J = 7.8 Hz) and a two proton-doublet at 8.19 ppm (J = 7.8 Hz), provide valuable information. First, the aromatic ring is symmetrically substituted since two protons are equivalent. Second, the high chemical shifts indicate that there are strongly deshielding atoms or groups on the aromatic ring. Finally, the splitting pattern tells us that the equivalent protons are next to the other set of protons. Therefore, we have a trisubstituted benzene ring or, more probably, a 2,6-disubstituted pyridine.

TABLE 24.1.S Analysis of the multiplets observed in the ^{1}H NMR spectrum of unknown compound **24**.

δ (ppm)	Integration	Multiplet	J (Hz)
0.92	6 H	Doublet	6.7
1.03	6 H	Doublet	6.9
1.85	2 H	Octect	6.7
4.13	2 H	Doublet of doublet of doublets	9.6, 8.4, 6.5
4.21	2 H	Triplet	8.5
4.51	2 H	Doublet of doublets	9.8, 8.4
7.84	1 H	Triplet	7.8
8.19	2 H	Doublet	7.8

The presence of two identical substituents is supported by the two six-proton doublets at 0.92 and 1.03 ppm (J = 6.7 and 6.8 Hz), with each corresponding to the resonance of two chemically equivalent methyl groups. The two-proton apparent octet at 1.85 ppm (J = 6.7 Hz) suggests two equivalent isopropyl fragments, which is a hypothesis that is quite probable, bearing in mind the base peak of the EI-MS spectrum as a result of the loss of 43 mass units (M-43). The multiplets at 4.13, 4.21, and 4.51 ppm are clearly the signals of non-equivalent protons, and taking into account that all of them integrate for two protons, we easily deduce that they are part of the twin substituent system. In addition, the resonance frequencies indicate that there are strong deshielding groups, probably heteroatoms, in their chemical environment.

From the rule of thirteen, considering that there is more than one heteroatom and 23 protons in the molecule, we obtain $C_{17}H_{23}N_3O_2$ and $C_{18}H_{23}NO_3$ among the possible molecular formulas. There are 9 signals in the ^{13}C NMR spectrum, but one of them is not duplicated, certainly corresponding to the carbon bound to the only proton that resonates at 7.84 ppm. Therefore, the former ($C_{17}H_{23}N_3O_2$) is the correct formula, which means that the compound has eight double-bond equivalents.

From the ^{1}H-^{13}C HSQC spectrum, we know that the aliphatic part of the twin substituent is composed of two methyl groups (18.4 and 19.2 ppm) and the CH (33.0) of the isopropyl fragment, one methylene (71.1 ppm) whose diastereotopic protons resonate at 4.21 and 4.51 ppm, and a second CH group (73.0 ppm). On the other hand, we can tell that two sp^2 carbons are bound to hydrogen, the ones at 125.8 and 137.3 ppm, and two are fully substituted, those resonating at 147.0 and 162.3 ppm.

The connectivity of the aromatic protons and the aliphatic spin system is easily obtained from the ^{1}H-^{1}H COSY spectrum (Fig. 24.1.S). As mentioned before, the protons of the latter system that resonate around

4.00 ppm must be next to heteroatoms, oxygen or nitrogen, to have such a high chemical shift. Since there are no more protons in the molecule, the twin substituents are probably cyclic in order to locate oxygen or nitrogen atoms within them.

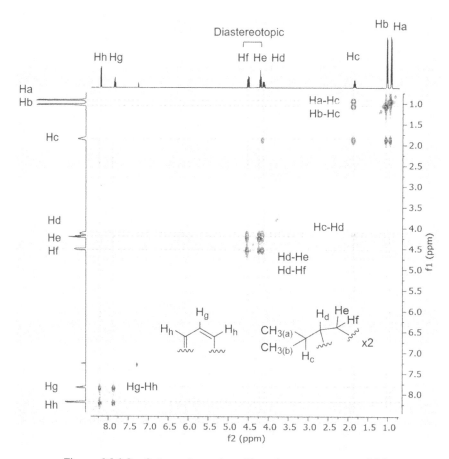

Figure 24.1.S   Spin systems found in unknown compound **24**.

In order to connect the pieces and propose a plausible structure, we have to consider that:

1. There are only three protons on the aromatic ring and they are very deshielded. Besides, we know that there is a duplicated substituent, so the most probable option is a 2,6-disubstituted pyridine.
2. The compound has eight unsaturations and only four of them are in the aromatic ring, so the other four must be distributed in the substituents, which probably will be cyclic.

3. The molecule has three nitrogen and two oxygen atoms, which means that we have to locate one nitrogen and one oxygen in each of the substituents in the case of a disubstituted pyridine, or all of the heteroatoms in the substituents of a hypothetical trisubstituted benzene ring.

4. There are two non-equivalent carbons that appear far downfield. In the case of a disubstituted pyridine, one of them obviously should be in the substituent.

Considering all of this data, we can readily deduce that the compound must be a 2,6-disubstituted pyridine having an isopropyldihydrooxazol or isopropyldihydroisoxazol as a substituent (Fig. 24.2.S). However, only in the oxazole type derivatives (I and II), the methylene and the adjacent CH would suffer the strong deshielding effect of a heteroatom that would cause the observed chemical shifts. Moreover, only in the 4-isopropyl isomer (I) we would observe both carbons around 70 ppm.

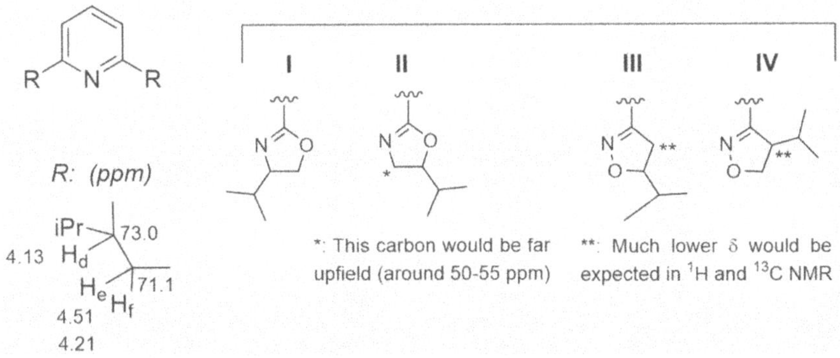

R: (ppm)

iPr 73.0
4.13 H$_d$
H$_e$ 71.1
4.51 H$_f$
4.21

*: This carbon would be far upfield (around 50-55 ppm)

**: Much lower δ would be expected in ^{1}H and ^{13}C NMR

Figure 24.2.S Possible structures suggested for unknown compound **24**.

On the other hand, the ^{1}H-^{13}C HMBC spectrum shows strong correlations between the quaternary carbon at 162.3 ppm and both the diastereotopic protons H$_e$ (4.21 ppm) and H$_f$ (4.51 ppm) and the proton of the carbon bound to the isopropyl group (H$_d$), thus suggesting similar couplings for all of them, which is something expected in the case of the oxazole cycle where all the correlations would be due to ^{3}J couplings (Fig. 24.3.S). It is worth mentioning that, although weak, two ^{4}J correlations (that of the quaternary carbon at 162.3 ppm with H$_g$, and the correlation of the other quaternary carbon at 147.0 ppm with H$_d$) are observed in the HMBC spectrum, probably due to the conjugation of the system and the angular disposition of the bonds.

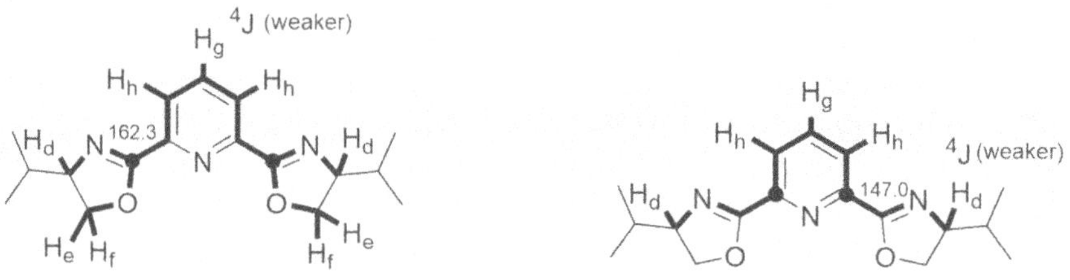

Figure 24.3.S   ¹H-¹³C long-range correlations from quaternary carbons observed in the HMBC spectrum of unknown compound **24**.

As a result of the above considerations, the assignment of the NMR signals for the proposed compound, 2,6-bis(4-isopropyl-4,5-dihydrooxazol-2-yl)pyridine, is shown below (Fig. 24.4.S).

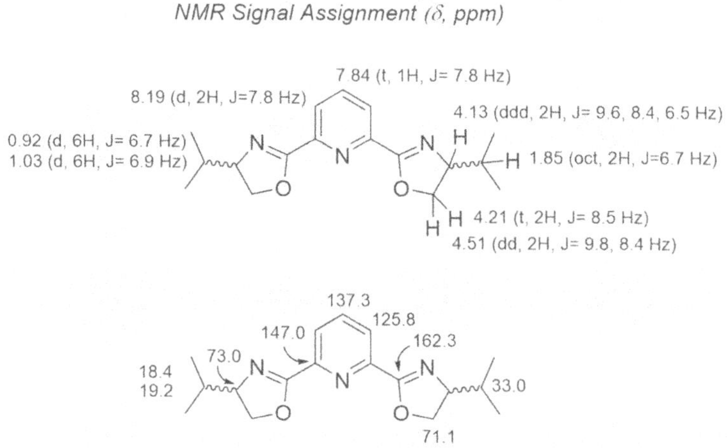

NMR Signal Assignment (δ, ppm)

Figure 24.4.S   NMR signal assignment of 2,6-bis(4-isopropyl-4,5-dihydrooxazol-2-yl)pyridine.

In addition to the loss of isopropyl and methyl radicals (M-43 and M-15 peaks), the fragmentation pattern of 2, 6-bis(4-isopropyl-4,5-dihydrooxazol-2-yl)pyridine is defined by several characteristic fragmentations of the 1, 3-oxazoline unit, based on the loss of neutral molecules, such as alkenes and imines, as well as the formation of the corresponding nitriles (Chen et al. 2005). Figure 24.5.S shows these fragmentations with a focus on those based on relatively complex rearrangements.

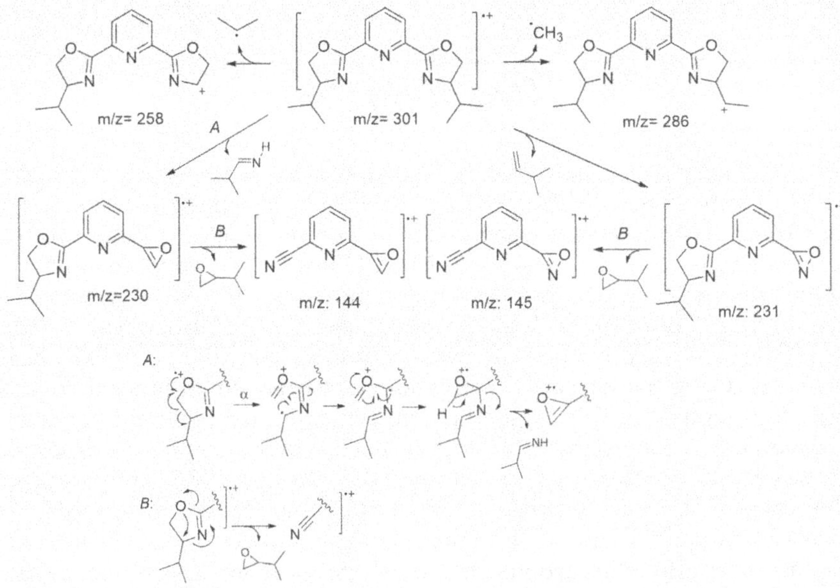

Figure 24.5.S Main mass fragmentations of 2,6-bis(4-isopropyl-4,5-dihydroox-azol-2-yl)pyridine.

Problem 25

The analysis of the EI-MS provides the m/z of the molecular ion (180) and some recognizable fragmentations, such as M-30 (typically loss of formaldehyde, m/z 150), M-43, and M-57. However, although an aliphatic ketone might fit some of the above fragmentations, there is no evidence of carbonyl carbons in the ^{13}C NMR spectrum. Besides, it shows four aromatic or alkene carbons, two of which are very deshielded (152.8 and 154.3 ppm) probably due to the effect of directly bound oxygen atoms. Since there are only four signals in this region, a symmetrically disubstituted benzene, a 5-membered heteroaromatic ring, or even a diene could be proposed. However, the intensity of the signals at 114.8 and 115.9 ppm suggest that each of them correspond to a pair of chemically equivalent CH, so a disubstituted benzene is the most plausible option. Apart from these signals, four aliphatic carbons with high chemical shifts (45-70 ppm) are observed, thus evidencing that they also suffer the deshielding effect of the aromatic ring and/or the oxygen atoms of the molecule.

The ^{1}H NMR spectrum confirms the conclusions drawn from ^{13}C NMR. The four aromatic proton-multiplet (6.8-6.9 ppm), which resembles two AB doublets with some small extra lines, supports the theory of a *para* disubstituted benzene theory because the observed coupling pattern is usually found in AA'BB' aromatic systems. Regarding the aliphatic moiety (see Table 25.1.S), the three-proton singlet reveals the presence of a methoxy group at 3.77 ppm, and the five one-proton multiplets between 2.7 and 4.2 ppm evidence that the rest of aliphatic protons are chemically non-equivalent, which means that there is a stereogenic center in the compound. The diverse coupling constants, away from the average values of vicinal constants in chains with free rotation, also suggest that the fragment is conformationally fixed. Except for the probable geminal coupling constant of the protons at 4.91 and 4.16 ppm (J = 11.1 Hz), it should be pointed out that the magnitude of the observed couplings is not too large.

TABLE 25.1.S Aliphatic signals in the ^{1}H NMR spectrum of unknown compound **25**.

δ (ppm)	Integration	Multiplet	J (Hz)
2.74	1 H	Doublet of doublets	5.0, 2.7
2.89	1 H	Triplet	4.6
3.32-3.35	1 H	Multiplet	–
3.77	3 H	Singlet	–
3.91	1 H	Doublet of doublets	11.1, 3.3
4.16	1 H	Doublet of doublets	11.1, 5.6

On the basis of the information so far, and applying the rule of thirteen, a molecular formula of $C_{10}H_{14}O_3$ with a degree of unsaturation of five is obtained. Accordingly, it is easily deduced that there is another cycle in the molecule besides the benzene ring - an extra cycle that would explain the observed ^1H-^1H couplings.

The ^1H-^{13}C HSQC spectrum confirms the 1,4-disubstituted aromatic ring and reveals the presence of two pairs of diastereotopic protons in the molecule, that is, those at 2.74 ppm and 2.89 ppm, bound to the carbon at 44.9 ppm, and the ones resonating at 4.16 and 3.91 ppm attached to the carbon at 69.7 ppm. Taking into account this information, the aliphatic spin system can be analyzed by using the ^1H-^1H COSY spectrum (Fig. 25.1.S).

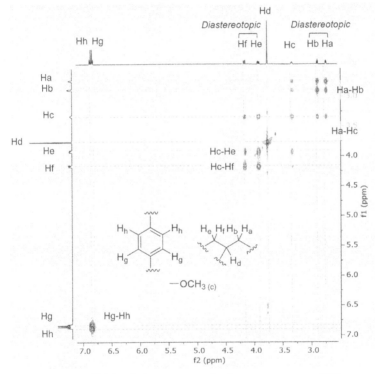

Figure 25.1.S Spin systems of unknown compound **25**.

As mentioned before, the quaternary carbons in the benzene ring are probably bound to oxygen atoms on account of their high chemical shift. One of the substituents is evidently the OMe group, so the aliphatic spin system is part of the second substituent. The chemical shifts of the protons H_e and H_f (3.91 and 4.16 ppm) and the corresponding carbon (69.7 ppm) indicate that the end of the aliphatic spin system is bound to the aryloxy fragment. Hence, the remaining signals must be the proton resonances of a monosubstituted oxirane ring. The observed coupling constants are in agreement with the proposed cycle. In fact, it is known that the homonuclear couplings in this kind of compounds are relatively small, with J_{cis} always being greater than J_{trans} ($J_{gem} \sim 6$ Hz, $J_{cis} \sim 5$ Hz and $J_{trans} \sim 3$ Hz).

With the structure of 2-((4-methoxyphenoxy)methyl) oxirane in mind, ^{1}H-^{13}C HMBC and ^{1}H-^{1}H NOESY spectra show the expected correlations that provide the required information for the assignment of the NMR signals (Fig. 25.2.S).

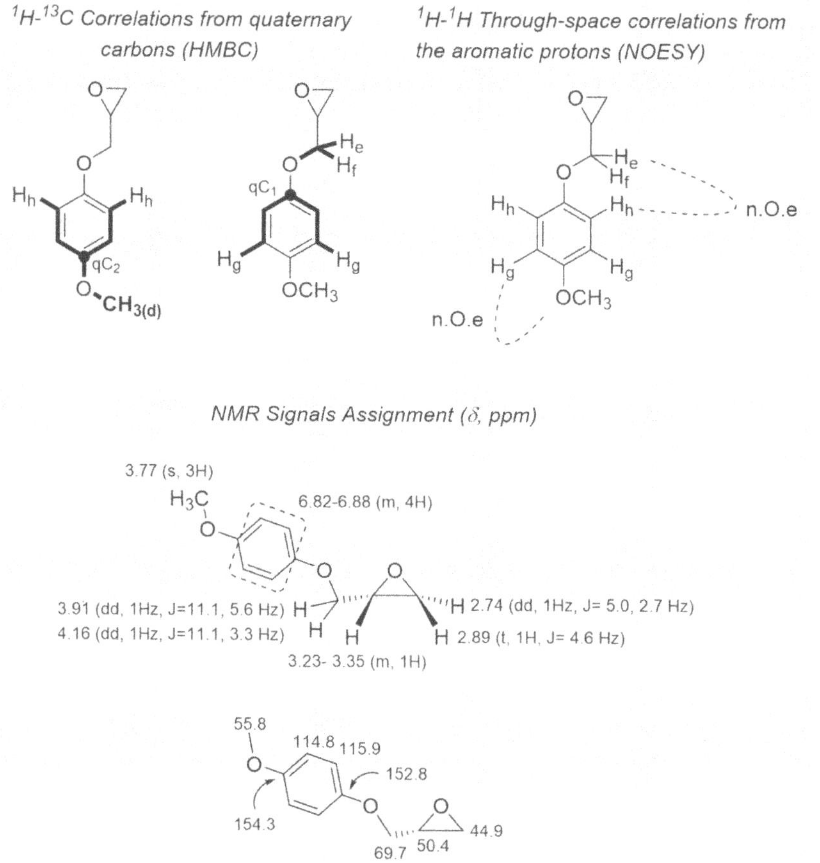

^{1}H-^{13}C *Correlations from quaternary carbons (HMBC)*

^{1}H-^{1}H *Through-space correlations from the aromatic protons (NOESY)*

NMR Signals Assignment (δ, ppm)

Figure 25.2.S   Key correlations observed from HMBC and NOESY spectra and NMR signal assignment.

Finally, the presence of the epoxide and methoxy moieties explains the M-30 (m/z 150), M-43 (m/z 137), and M-57 peaks (m/z 123) of the mass spectrum. Indeed, these peaks are due to the loss of formaldehyde and oxiranyl and oxiranylmethyl radicals, respectively. The above cleavages are followed by decarbonylations, and the loss of other neutral molecules, such as acetylene, formaldehyde, or cetene. For a more thorough analysis of the fragmentation pattern of 2-((4-methoxyphenoxy)methyl)oxirane, see Figure 25.3.S.

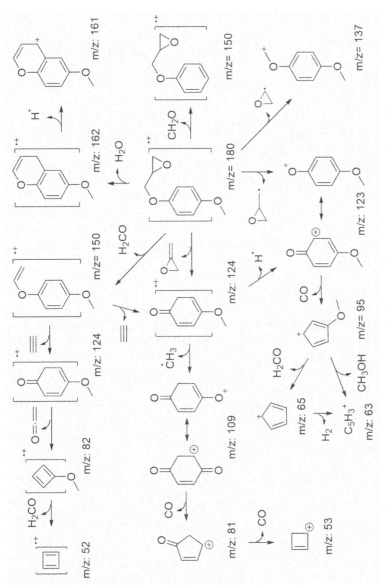

Figure 25.3.S Mass fragmentation pattern of 2-((4-methoxyphenoxy)methyl)oxirane.

Problem 26

The odd molecular mass indicates that we are dealing with the structure of a nitrogen compound. Thanks to the accurate mass, its molecular formula, $C_{16}H_{13}NO$, can be easily calculated, revealing that the compound contains 11 double bond equivalents.

The infrared spectrum shows an intense absorption at 1639 cm^{-1} that could correspond to the stretching of a double bond or even to the carbonyl group of a conjugated amide, specifically a tertiary amide, since no absorption of NH bonds is present in the spectrum. Two weaker bands at 1610 and 1588 cm^{-1} suggest the presence of an aromatic ring ($v_{C_{ar}-C_{ar}}$), a fact which is supported by the presence of C_{sp^2}-H stretching bands at 3073 and 3034 cm^{-1}. In addition, the molecule probably has an aliphatic moiety on account of the $v_{C_{sp^3}-H}$ bands found at 2940 and 2872 cm^{-1}.

The 16 signals of the ^{13}C NMR spectrum reveal the absence of symmetry elements in the molecule. As the high unsaturation index suggested, many of the carbon atoms are Csp^2. In view of the 12 resonances between 119.3 and 134.7 ppm, there must be more than one aromatic ring in the structure - probably two benzene rings. Besides, the downfield resonance at 161.3 ppm lends credence to the idea of an amide function in the molecule. The remaining carbons are clearly sp^3 since they resonate further upfield, exactly at 20.9, 28.4, and 42.9 ppm. On account of the relatively high chemical shift of the last one, a deshielding effect by the amide nitrogen can be suggested. Two benzene rings and one carbonyl group would account for nine unsaturations, so the compound should contain two additional rings.

TABLE 26.1.S ^{1}H NMR signals of unknown compound **26**.

δ (ppm)	Integration	Multiplet	J (Hz)
2.07	2 H	Quintet	6.1
2.94	2 H	Triplet	6.2
4.21-4.27	2 H	Multiplet	–
7.13	1 H	Triplet	7.5
7.22	1 H	Doublet of doublets	7.3, 1.1
7.50	1 H	Doublet of doublet of doublets	8.1, 7.1, 1.1
7.67	1 H	Doublet of doublet of doublets	8.3, 7.2, 1.4
8.05	1 H	Doublet of doublets	7.9, 1.1
8.19	1 H	Doublet	8.1
8.47	1 H	Doublet of doublets	8.0, 1.1

The ¹H NMR spectrum shows seven one-proton multiplets in the aromatic region, whose splitting patterns are consistent with 1,2-disubstituted and 1,2,3-trisubstituted benzene rings (Table 26.1.S). Regarding the chemical shifts, the fact that three signals appear above 8 ppm might suggest a pyridine ring, but there is no evidence of such heterocycle in the ¹³C NMR spectrum (no signals around 150 ppm). The rest of the protons give rise to three two-proton multiplets centered at 2.07, 2.94, and 4.24 ppm, thus suggesting a separate aliphatic spin system predictably bound to the amide nitrogen and one of the aromatic rings, on account of the observed chemical shifts.

The ¹H-¹³C HSQC spectrum allows an easy identification of the aromatic carbons directly attached to protons and, consequently, the five fully substituted ones (those at 119.3, 125.5, 125.7, 133.8, and 134.7 ppm) which are essential to determine the structure of polycyclic compounds. The aliphatic carbons, as expected, are three methylene groups correlated to the two-proton multiplets of the ¹H NMR.

The ¹H-¹H COSY spectrum confirms the three spin systems (Fig. 26.1.S). In fact, the absence of correlations due to *meta* couplings in this case enables a complete analysis of the disubstituted and trisubstituted aromatic spin systems.

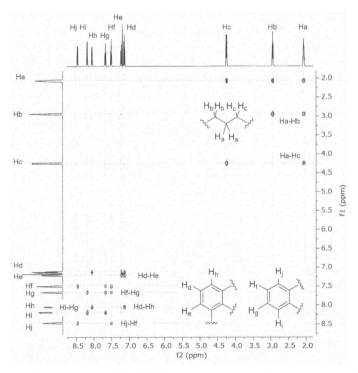

Figure 26.1.S Analysis of the ¹H-¹H spin systems of unknown compound **26**.

We just need to put together the pieces of the puzzle already deduced (Fig. 26.2.S). It is evident that, as the unsaturation index suggested, the formation of two additional cycles is necessary to get a plausible structure. The study of the long-distance couplings between carbon-13 nuclei and protons in the ^1H-^{13}C HMBC spectrum should provide the required information.

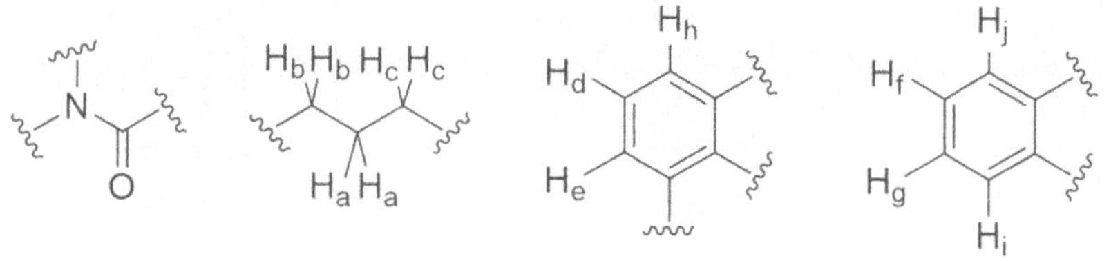

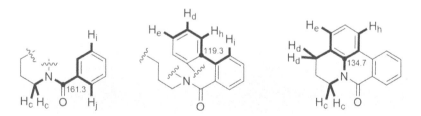

Figure 26.2.S   Structural fragments of unknown compound **26**.

In the ^1H-^{13}C HMBC spectrum, the quaternary carbon at 161.3 ppm, that is, the amide carbonyl carbon, correlates to the protons of the aliphatic spin system that resonate at 4.2 ppm (H_c). Therefore, those methylene protons are attached to the carbon bound to the amide nitrogen, as expected (Fig. 26.3.S). That carbonyl carbon is also correlated with the protons at 8.47 ppm (H_j) and 8.19 ppm (H_i). Hence, we know that the amide moiety is one of the two substituents of disubstituted benzene ring. The other substituent is clearly the second benzene ring, since the quaternary carbon at 119.3 ppm correlates to the protons of the three-proton aromatic spin system (H_e, H_d, and H_c) and to H_i. The analogous intense correlations between that quaternary carbon and the H_d and H_i protons suggest a three bond-coupling in both cases. Finally, the correlations of the aromatic carbon attached to the amide nitrogen (134.7 ppm) confirm the identity of the third substituent of the trisubstituted aromatic ring, at the other end of the aliphatic spin system (H_d). We have already solved the problem. The unknown compound is 5,6-dihydro-4H,8H-pyrido[3,2,1-de] phenanthridin-8-one.

Figure 26.3.S   Key correlations observed in ^1H-^{13}C HMBC.

Accordingly, after a careful analysis of all the correlations shown by ¹H-¹³C HMBC and HSQC spectra, the full assignment of the NMR signals is shown in Figure 26.4.S.

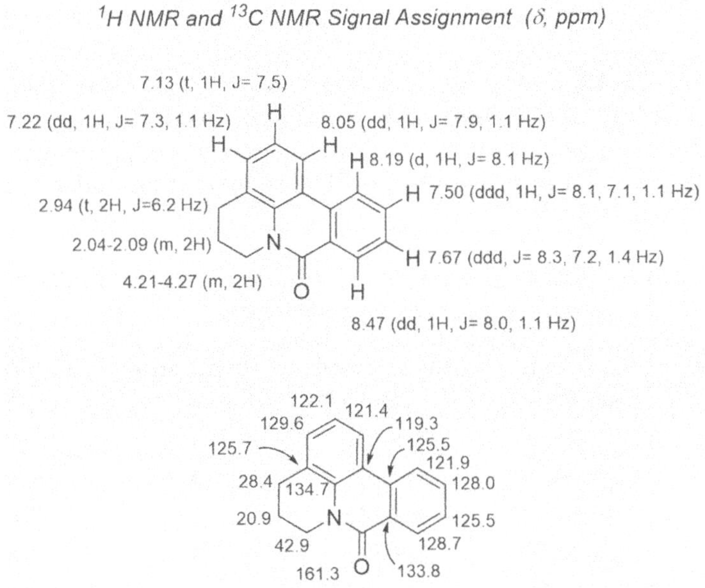

¹H NMR and ¹³C NMR Signal Assignment (δ, ppm)

Figure 26.4.S Assignment of the NMR signals of 5,6-dihydro-4H,8H-pyr-ido[3,2,1-de]phenanthridin-8-one.

Problem 27

The even m/z value of MH$^+$ ion corresponds to a nitrogen compound. In this case, the online tool (the molecular formula calculator from monoisotopic mass of the webpage www.cheminfo.org has been used) provides a list of possible molecular formulas (Table 27.1.S), so more information is needed in order to narrow this range.

TABLE 27.1.S Possible molecular formulas for MH$^+$ 346.1106 from the online tool at www.cheminfo.org.

Entry	Molecular Formula	Monoisotopic Mass	Unsaturation	Error (ppm)
1	$C_{16}H_{21}F_2NOS_2$	345.1032	6	−0.18
2	$C_{18}H_{19}NO_4S$	345.1035	10	0.45
3	$C_{12}H_{18}F_3NO_7$	345.1035	3	0.62
4	$C_{20}H_{16}N_3OP$	345.1031	15	−0.65
5	$C_9H_{21}FN_5O_4PS$	345.1036	2	0.77

No distinguishable bands for amide C=O or amide NH bonds are found in the IR spectrum. The intense absorption at 1768 cm^{-1} probably corresponds to the carbonyl stretch of an ester group. Besides, the bands at 1491 and 1594 cm^{-1} suggest the presence of an aromatic ring ($\nu_{C_{ar}-C_{ar}}$), which is a proposal supported by the weak C_{sp^2}-H stretching band at 3062 cm^{-1}. Two weak ν_{Csp^3}-H bands at 2979 and 2878 cm^{-1} indicate the presence of an alkyl fragment. Although within the fingerprint region and therefore not reliable, the relatively strong bands at 1343 and 1159 cm^{-1} might be due to the stretching of a SO_2N unit, thus supporting the idea of a molecular formula containing sulfur.

The information extracted from the IR spectrum is corroborated by the $^{13}C\{^1H\}$ NMR spectrum, which shows the resonance of a carbonyl carbon at 171 ppm and eight aromatic carbons between 121 and 150 ppm. Thanks to DEPT-135 spectrum, we know that three of those aromatic carbons are fully substituted. The quite high chemical shifts of two of them suggest that they are bound to electron-withdrawing groups, probably heteroatoms, at least in the case of the resonance at 150.5 ppm. The molecule also has an aliphatic moiety, as the upfield region of the spectrum clearly shows. There are five signals in this zone, one at 21.6 ppm, predictably due to a methyl group, three non-equivalent methylene carbons at 24.8, 31.1, and 48.65 ppm, and a probable methine group at 60.5 ppm, shifted downfield as a result of the presence of an oxygen atom.

With regards to the molecular formula, there is no trace of multiplets in the $^{13}C\{^1H\}$ NMR spectrum, which makes the presence of fluorine or phosphorus atoms in the molecule unlikely. Thus, the option of the nitrogen-sulfur compound becomes the most plausible ($C_{18}H_{19}NO_4S$, Table 27.1.S, entry 2). The number of signals (14) that show up in the $^{13}C\{^1H\}$ NMR spectrum is understandable, considering that due to symmetry, some carbon nuclei may be chemically equivalent. It is important to remember that in this case, caution should be taken when considering the unsaturation index of the compound, since the oxidation state of sulfur in the molecule is unknown. This leads us to the previous hypothesis on the appearance in the infrared spectrum of the SO_2 stretching bands of a sulfonamide, thus meaning that the double bound equivalents would be 12 and not 10.

The ^1H NMR spectrum confirms that there are 19 protons in the molecule. These protons give rise to 12 signals, eleven of which are multiplets whose integration and multiplicity are shown in Table 27.2.S. The number and complexity of the signals due to the aliphatic protons, except for the three-proton singlet of a methyl group at 2.34 ppm, indicate the presence of a spin system in the molecule formed by seven strongly coupled non-equivalent protons. The observed multiplicity and chemical shifts suggest that the protons at 4.40, 3.47, and 3.29 ppm are located at both ends of the alkyl chain and next to deshielding groups. The aromatic signals reveal clearly a *para* disubstituted benzene, whose protons resonate at 7.24 and 7.73 ppm, and an additional monosubstituted benzene

TABLE 27.2.S ^1H NMR signals of unknown compound **27**.

δ (ppm)	Integration	Multiplet	J (Hz)
1.95-2.03	2H	Multiplet	–
2.09-2.15	2H	Multiplet	–
2.34	3H	Singlet	–
3.29	1H	Doublet of triplets	9.5, 7.2
3.47	1H	Doublet of doublet of doublets	9.5, 7.4, 4.9
4.40	1H	Doublet of doublets	7.3, 5.7
7.00-7.05	2H	Multiplet	–
7.13-7.18	1H	Multiplet	–
7.24	2H	Doublet	8.0
7.28-7.33	2H	Multiplet	–
7.73	2H	Doublet	8.2

ring with two two-proton multiplets and one-proton multiplet which is difficult to resolve. The two arenes and the observed ester group would account for 9 unsaturations (11 if we consider the probable SO_2 group), so we should consider an additional ring in order to satisfy the unsaturation degree and to connect the pieces deduced so far.

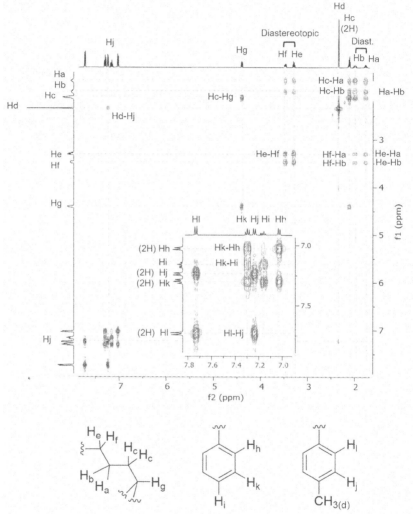

Figure 27.1.S Spin systems deduced from ¹H-¹H COSY of unknown compound **27**.

From the ¹H-¹³C HSQC spectrum, the signals of the diastereotopic aliphatic protons are easily identified. The resonances at 3.47 and 3.29 ppm are due to the protons attached to the carbon at 48.5 ppm, the multiplets centered at 2.00 ppm, and 1.75 ppm are the signals of those attached to the carbon at 24.7 ppm, and the two-proton multiplet at 2.1 ppm is the signal of the protons of the carbon that resonates at 31.0 ppm. Logically, the signal at 4.40 ppm is correlated to the aliphatic methine carbon at 60.5 ppm. Besides, this two-dimensional experiment easily shows that the signals at 121.3, 127.6, 129.4, and 129.7 ppm are the resonances of four pairs of equivalent CH carbons, as expected from the above hypothesized *para*-disubstituted and a monosubstituted benzene rings. However, it is impossible to distinguish between the protons attached to the carbons at 129.4 and 129.7 ppm because of the proximity of the signals.

The aforementioned three spin systems are easily analyzed by ^{1}H-^{1}H COSY (Fig. 27.1.S). As ventured, the most deshielded aliphatic protons are at the ends of the aliphatic chain and clearly close to the nitrogen, ester, or sulfonyl groups. An unexpected correlation due to weak coupling between the three proton singlet (H_d) and two of the protons of the disubstituted benzene ring (H_j) facilitates the positioning of the methyl group in the *para*-disubstituted benzene.

As in other examples, the clues to connect the spin systems are provided by the ^{1}H-^{13}C HMBC (Fig. 27.2.S). It can be observed that the quaternary carbon at 170.7 ppm, the ester carbonyl, is correlated to the aliphatic spin system, to be precise to H_g (4.40 ppm) and to H_c protons (2.09-2.15 ppm)[7]. This ester carbon does not show more correlations, but it is easy to deduce that it is next to the monosubstituted benzene fragment. Indeed, the quaternary carbon of that phenyl ring resonates at 150.5 ppm, as the HMBC experiment reveals (Fig. 27.2.S). The remaining aromatic signals can be assigned following the correlations of the two quaternary carbons of the *para*-disubstituted benzene ring. Besides, the lack of correlation between the nuclei of this aromatic ring and the other spin systems can be explained if a connection through a -N-SO$_2$- unit is considered. With these facts, the missing unsaturation certainly due to a cycle, and the high chemical shift of the aliphatic methine hydrogen in mind, it is simple to deduce that the unknown compound **27** is phenyl *N*-tosylprolynate.

The ^{1}H-^{1}H NOESY experiment supports this proposal (Fig. 27.2.S). Thus, in addition to the expected through-space dipolar couplings within the pyrrolidine ring and between aromatic protons in each arene unit, the spectrum shows clear correlations between the protons of the tosyl group resonating at 7.76 ppm (H_l) and the protons at the two closest positions of the pyrrolidine ring, i.e., H_g (4.40 ppm) and H_e and H_f (3.29 and 3.47 ppm).

The topological information extracted from 2D NMR correlations facilitates the assignment of the NMR signals save for the aforementioned CH carbons at 129.4 and 129.7 ppm (Fig. 27.3.S).

---

[7] Although H_c protons are chemically non-equivalent, they are referred as such because their signals appear overlapped.

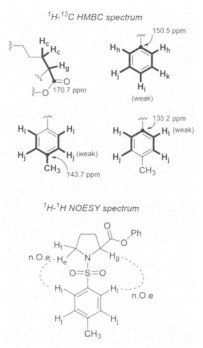

Figure 27.2.S   Key correlations observed in ^1H-^{13}C HMBC and ^1H-^1H NOESY spectra.

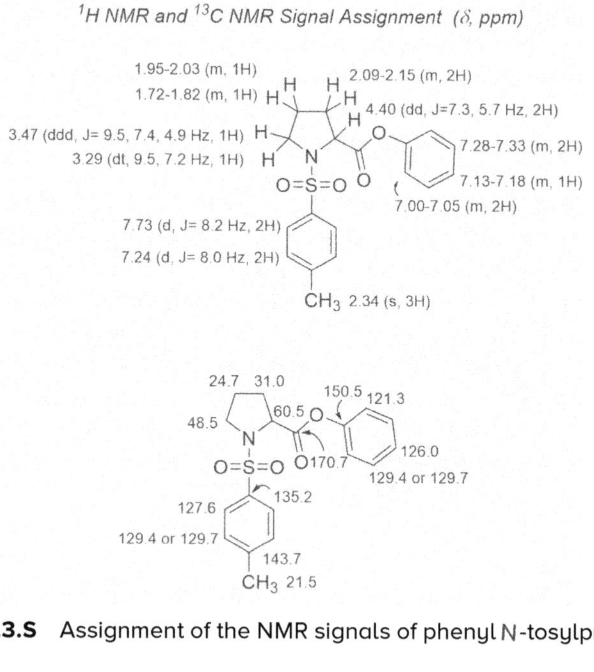

^1H NMR and ^{13}C NMR Signal Assignment (δ, ppm)

1.95-2.03 (m, 1H)
1.72-1.82 (m, 1H)
2.09-2.15 (m, 2H)
4.40 (dd, J=7.3, 5.7 Hz, 2H)
3.47 (ddd, J= 9.5, 7.4, 4.9 Hz, 1H)
3.29 (dt, 9.5, 7.2 Hz, 1H)
7.28-7.33 (m, 2H)
7.13-7.18 (m, 1H)
7.00-7.05 (m, 2H)
7.73 (d, J= 8.2 Hz, 2H)
7.24 (d, J= 8.0 Hz, 2H)
CH₃ 2.34 (s, 3H)

24.7   31.0
48.5
60.5
150.5   121.3
126.0
170.7
129.4 or 129.7
135.2
127.6
129.4 or 129.7
143.7
CH₃ 21.5

Figure 27.3.S   Assignment of the NMR signals of phenyl N-tosylprolynate.

References

Abraham, J.R. and H.J. Berstein. 1959. The proton resonance spectra of furan and pyrrole. Can. J. Chem. 37: 1056-1065.

Abraham, J.R. and H.J. Berstein. 1961. The proton resonance spectra of some substituted furans and pyrroles. Can. J. Chem. 39: 905-914.

Aires-de-Sousa, J., M.C. Hemmer and J. Gasteiger. 2002. Prediction of ^{1}H NMR chemical shifts using neural networks. Anal. Chem. 74: 80-90.

Aydin, R. and H. Günther. 1990. ^{13}C, ^{1}H spin-spin coupling. X.-Norbornane: A reinvestigation of the Karplus curve for $^3J_{(13C-1H)}$. Magn. Reson. Chem. 28: 448-457.

Balacco, G. 1996. A desktop calculator for the Karplus equation. J. Chem. Inf. Comput. Sci. 36: 885-887.

Banfi, D. and L. Patiny. 2008. www.nmrdb.org: Resurrecting and processing NMR spectra on-line. Chimia. 62: 280-281.

Binev, M., M. Marques and J. Aires-de-Sousa. 2007. Prediction of ^{1}H NMR coupling constants with associative neural networks trained for chemical shifts. J. Chem. Inf. Model. 47: 2089-2097.

Bothner-By, A.A. 1965. Geminal and vicinal proton-proton coupling constants in organic compounds. Adv. Magn. Reson. 1: 195-316.

Bright, J.W. and E.C. Chen. 1983. Mass spectral interpretation using rule of 13. J. Chem. Educ. 60: 557-558.

Cahill, R., R.C. Cookson and T.A. Crabb. 1969. Geminal coupling constants in methylene groups—II: J in CH2 groups α to heteroatoms. Tetrahedron. 25: 4681-4709.

Castillo, A.M., L. Patiny and J. Wist. 2011. Fast and accurate algorithm for the simulation of NMR spectra of large spin systems. J. Magn. Res. 209: 123-130.

ChemBioDraw Ultra version 13.0.0.3015. 2012. PerkinElmer.

Chen, H., D.-M. Du, B. Fu and M.-Y. He. 2005. Investigation on mass spectral fragmentation mechanism of C_2-symmetric chiral bis(oxazoline) and bis(thiazoline). Chin. J. Chem. 23: 720-724.

Claridge, T.D.W. 2009. High Resolution NMR Techniques in Organic Chemistry. Tetrahedron Organic Chemistry Series. 27. Elsevier, Hungary.

Cookson, R.C., T.A. Crabb, J.J. Frankel and J. Hudec. 1966. Geminal coupling constants in methylene groups. Tetrahedron. 22: 355-390.

Davis, J.C. Jr. and T.V. Van Auken. 1965. Chemical shifts and spin-spin coupling interactions in the nuclear magnetic resonance spectra of endo- and exo-nor-bornene derivatives. J. Am. Chem. Soc. 87: 3900-3905.

Friebolin, H. 2010. One-and Two-Dimendsional NMR Spectroscopy. An Introduction. VCH, Weinheim.

Günther, H. and G. Jikeli. 1977. ^{1}H nuclear magnetic resonance spectra of cyclic monoenes: Hydrocarbons, ketones, heterocycles and benzo derivatives. Chem. Rev. 77: 599-637.

Günther, H. 1992. NMR Spectroscopy. Georg Thieme, Sttutgart, New York.

Haasnoot, C.A.G., F.A.A.M. DeLeeuw and C. Altona. 1980. The relationship between proton-proton NMR coupling constants and substituent electronegativities-I: An empirical generalization of the Karplus equation. Tetrahedron. 36: 2783-2792.

Hesse, H., H. Meier and B. Zeels. 2007. Spectroscopic Methods in Organic Chemistry. Thieme Publishing Group. Stuttgart.

Huckerby, T.N. 1970. General review of proton magnetic resonance. Annu. Rep. NMR Spectro. 3: 43.

Jacobsen, N.E. 2017. NMR Data Interpretation Explained. Wiley 2017, New Jersey. 182-183.

Karplus, M. 1963. Vicinal proton coupling in nuclear magnetic resonance. J. Am. Chem. Soc. 85: 2870-2871.

Levitt, M.H. 2001. Spin Dynamics: basics of nuclear magnetic resonance. J. Wiley-Sons, Chichester.

Levy, G.C., R.L. Lichter and G.L. Nelson. 1980. Carbon-13 Nuclear Magnetic Resonance Spectroscopy. Wiley-Interscience, New York.

Marshall, J.L. 1983. Carbon-Carbon and Carbon-Proton NMR Couplings: Applications to Organic Stereochemistry and Conformational Analysis. Verlag Chemie International, Deerfield Beach, FL.

Martin, N.H., N.W. Allen, E.K. Minga, S.T. Ingrassia and J.D. Brown. 1998. Computational evidence of NMR deshielding of protons over a carbon–carbon double bond. J. Am. Chem. Soc. 120: 11510-11511.

Naught, A.D. and A. Wilkinson. 1997. IUPAC. Compendium of Chemical Terminology. Blackwell Scientific Publications, Oxford.

Patiny, L. and A. Borel. 2013. ChemCalc: A building block for tomorrow's chemical infrastructure. J. Chem. Inf. Model. 535: 1223-1228.

Pavia, D.L., G.M. Lampman, G.S. Kriz and J.R. Vyvyan. 2009. Introduction to Spectroscopy. Brooks/Cole, Cengage Learning. 4th Ed.

Pretsch, E., P. Bühlmann and M. Badertscher. 2010. Structure Determination of Organic Compounds: Tables of Spectral Data. Springer, Berlin.

Reich, H.J. 2018. 5-HMR-5 Vicinal Proton-Proton Coupling $^3J_{HH}$. https://www.chem.wisc.edu/areas/reich/nmr/05-hmr-05-3j.htm (accessed June 3, 2019).

Silverstein, R.M., F.X. Webster and D.J. Kiemle. 2014. Spectrometric Identification of Organic Compounds. John Wiley & Sons: Hoboken, NJ.

Steinbeck, Ch., S. Krause and S. Kuhn. 2003. NMRShiftDB constructing a free chemical information system with open-source components. J. Chem. Inf. Comput. Sci. 43: 1733-1739.

Stévigny, C., J.-L. Habid Jiwan, R. Rozenberg, E. de Hoffmann and J. Quetin-Leclercq. 2004. Key fragmentation patterns of aporphine alkaloids by electrospray ionization with multistage mass spectrometry. Rapid Commun. Mass Spectrom. 18: 523-528.

Tori, K., Y. Hata, R. Muneyuki, Y. Takano, T. Tsuji and H. Tanida. 1964. N.M.R. studies of bridged ring systems: II. Unusual magnetic deshielding effect on the bridge methylenes of norbornadiene and benzonorbornadiene. Can. J. Chem. 42: 926-933.

Tvaroska, I. and F.R. Taravel. 1995. Carbon-proton coupling constants in the conformational analysis of sugar molecules. Adv. Carbohydrate Chem. Biochem. 51: 15-61.

Wasylishen, R. and T. Schafer. 1973. INDO molecular orbital calculations of nuclear spin–spin coupling constants over three bonds between ^{13}C and ^{1}H in some simple molecules. Can. J. Chem. 51: 961-973.

Williams, D. and I. Fleming. 2008. Spectroscopic Methods in Organic Chemistry. McGraw-Hill, London.

Index

Milton Keynes UK
Ingram Content Group UK Ltd.
UKHW040105071024
449327UK00019B/816